Thinking Like a Mathematician

WILLIAM & MARY'S CENTER FOR GIFTED EDUCATION

Thinking Like a Mathematician

GRADE 3

Lessons That Develop Habits of Mind and Thinking Skills for Young Mathematicians

**MARY-LYONS WALK HANKS,
JENNIFER K. LAMPERT, AND
KATHERINE PLUM**

CENTER FOR GIFTED EDUCATION

P.O. Box 8795
Williamsburg, VA 23187

Copyright ©2019, Prufrock Press Inc.

Edited by Stephanie McCauley

Cover design and layout design by Micah Benson

ISBN-13: 978-1-61821-824-7

No part of this book may be reproduced, translated, stored in a retrieval system, or transmitted, in any form or by any means, electronic, mechanical, photocopying, microfilming, recording, or otherwise, without written permission from the publisher.

Prufrock Press grants the individual purchasing this book permission to photocopy original activity pages for single classroom use. This permission does not include electronic reproduction rights. Should you wish to make copies of materials we sourced or licensed from others, request permission from the original publisher before reproducing that material.

For more information about our copyright policy or to request reprint permissions, visit https://www.prufrock.com/permissions.aspx.

Printed in the United States of America.

At the time of this book's publication, all facts and figures cited are the most current available. All telephone numbers, addresses, and website URLs are accurate and active. All publications, organizations, websites, and other resources exist as described in the book, and all have been verified. The authors and Prufrock Press Inc. make no warranty or guarantee concerning the information and materials given out by organizations or content found at websites, and we are not responsible for any changes that occur after this book's publication. If you find an error, please contact Prufrock Press Inc.

Prufrock Press Inc.
P.O. Box 8813
Waco, TX 76714-8813
Phone: (800) 998-2208
Fax: (800) 240-0333
http://www.prufrock.com

TABLE OF CONTENTS

Introduction **1**

Introductory Lesson: What Is a Mathematician? **6**

Unit 1: Systems of Counting **11**
    Lesson 1.1: Explore Different Counting Systems **12**
    Lesson 1.2: Apply Binary Counting **17**
    Lesson 1.3: Project: Systems of Counting Around the World ... **22**

Unit 2: Order of Operations **33**
    Lesson 2.1: Explore the Order of Operations **34**
    Lesson 2.2: Apply the Order of Operations **38**
    Lesson 2.3: Project: Reflect on the Order of Operations **43**

Unit 3: Patterns **49**
    Lesson 3.1: Explore Patterns **50**
    Lesson 3.2: Apply Patterns to the Golden Spiral **54**
    Lesson 3.3: Project: Analyze Patterns in Art **58**

Unit 4: Time and Measurement **65**
    Lesson 4.1: Explore Time on Planets **66**
    Lesson 4.2: Apply Solar Math **70**
    Lesson 4.3: Project: Design Solar Models **74**

Unit 5: Graphing, Data, and Charts in Algebra **79**
    Lesson 5.1: Explore Linear Data, Charts, and Graphs **80**
    Lesson 5.2: Apply Nonlinear Data, Charts, and Graphs **84**
    Lesson 5.3: Project: Create a Business Plan **89**

THINKING LIKE A MATHEMATICIAN

Unit 6: Geometry **97**
    Lesson 6.1: Explore Perimeter and Area **98**
    Lesson 6.2: Apply Properties of Quadrilaterals **103**
    Lesson 6.3: Project: Analyze Famous Architecture **108**

Unit 7: Data Analysis and Statistics **115**
    Lesson 7.1: Explore Tidal Data and Statistics **116**
    Lesson 7.2: Apply Data Analysis **120**
    Lesson 7.3: Project: Shipwrecked! **124**

Final Project: Designing a Garden **129**

Answer Key **135**

About the Authors **143**

INTRODUCTION

PURPOSE

Math is everywhere, and it is exciting to introduce students to real-world applications of the concepts they study in school each day. *Thinking Like a Mathematician* provides opportunities to connect your math content to diverse mathematicians, introductory activities, and real-world applications.

This book invites students to engage with seven units of study and a culminating problem-based assessment. There are multiple ways to use this text. The units can be taught sequentially, resulting in a thorough exploration of the field of mathematics, or educators can choose to teach the units, lessons, and projects in isolation to complement an existing curriculum. We hope that *Thinking Like a Mathematician* guides your students' learning in rich and meaningful ways.

ORGANIZATION OF THE BOOK

Lesson 1 (Explore) and Lesson 2 (Apply) of each unit are designed for you to instruct students on new material. These activities might be most meaningful for students who have a little background knowledge prior to beginning, but they are certainly full of teachable moments for you to guide student learning. Each lesson's activities require students to think about math in different ways, but please reflect on your individual students' needs as you make adjustments.

Each unit concludes with a research project in Lesson 3. Students could investigate each topic as partners or as individuals in these highly adaptable projects. The projects provide students structured guidance but also allow choices on what topic to explore. As students learn to research, you should guide their time management. Partnerships with your school librarian and technology staff will help build students' research skills. The rubrics for the projects are included at the end of each unit.

As students begin to work in groups, they will need your direct instruction on the math components, but also the components required to be helpful, contributing teammates. These interpersonal skills are the foundation for the projects in this book.

STUDENT MATH NOTEBOOKS

Writing and reflecting on math concepts are important skills for students to develop. Many of the lessons make reference to student math notebooks, in which students can answer Anticipatory Set and Exit Ticket questions. Teachers should provide spirals or binders for students to use. Students should also answer questions in the math notebook if there is not a handout to use. To make these notebooks more dynamic, students may pick pictures of mathematicians (and themselves) for the notebook cover. Students can build these math notebooks to use as a reference throughout the year.

REFLECTION

As they explore content and interpersonal relationships, students need time to reflect upon their progress. They should explore their background knowledge and new learning, consider which products might represent their learning best, decide how to analyze and tackle obstacles, and learn about their own time management skills. Included with Lesson 1.3 is a reflection page, which is designed for students to visit multiple times throughout each of the projects, not just once at the end. Learning time is precious, and a quick conversation can be more meaningful than a lengthy analysis.

The student reflection helps students consider their effort, teamwork, and product. High-ability students may be used to coasting through assigned classwork independently. These rigorous projects require them to investigate real-world problems in challenging ways. Inevitably, students will confront obstacles, allowing for discussion and teachable moments.

STANDARDS

The units are organized around the content standards from the National Council of Teachers of Mathematics (NCTM). We have written each unit to correspond to one of the objectives (see Table 1).

Introduction

Table 1
NCTM Standards Alignment

Unit	NCTM Content Standard
Unit 1	**Numbers and Operations:** • Understand numbers, ways of representing numbers, relationships among numbers, and number systems • Understand meanings of operations and how they relate to one another • Compute fluently and make reasonable estimates
Unit 2	**Numbers and Operations:** • Understand numbers, ways of representing numbers, relationships among numbers, and number systems • Understand meanings of operations and how they relate to one another • Compute fluently and make reasonable estimates
Unit 3	**Algebra** • Understand patterns, relations, and functions • Represent and analyze mathematical situations and structures using algebraic symbols • Use mathematical models to represent and understand quantitative relationships • Analyze change in various concepts
Unit 4	**Measurement** • Understand measurable attributes of objects and the units, systems, and processes of measurement • Apply appropriate techniques, tools, and formulas to determine measurements
Unit 5	**Algebra** • Understand patterns, relations, and functions • Represent and analyze mathematical situations and structures using algebraic symbols • Use mathematical models to represent and understand quantitative relationships • Analyze change in various concepts
Unit 6	**Geometry** • Analyze characteristics and properties of two- and three-dimensional shapes and develop mathematical arguments about geometric relationships • Specify locations and describe spatial relationships using coordinate geometry and other representational systems • Apply transformations and use symmetry to analyze mathematical situations • Use visualization, spatial reasoning, and geometric modeling to solve problems

Table 1, continued

Unit	NCTM Content Standard
Unit 7	**Geometry** • Analyze characteristics and properties of two- and three-dimensional shapes and develop mathematical arguments about geometric relationships • Specify locations and describe spatial relationships using coordinate geometry and other representational systems • Apply transformations and use symmetry to analyze mathematical situations • Use visualization, spatial reasoning, and geometric modeling to solve problems
Unit 8	**Data Analysis & Probability** • Formulate questions that can be addressed with data and collect, organize, and display relevant data to answer them • Select and use appropriate statistical methods to analyze data • Develop and evaluate inferences and predictions that are based on data • Understand and apply basic concepts of probability
Final Project	Standards used will vary based on individual student projects.

SUPPLEMENTAL BIOGRAPHIES

We have also developed a set of biographies of famous mathematicians that can be accessed at https://www.prufrock.com/thinking-like-a-mathematician.aspx. Each biography's content is linked to one of the units. Asking students to think like a mathematician requires an awareness of the many facets of the field. The biographies highlight both men and women across centuries and cultures, and have a readability geared for advanced third graders. The biography passages contain rich vocabulary and invite an interdisciplinary approach to math. These are excellent opportunities to teach students to read critically for details and apply reading strategies.

You may implement these biographies as an introduction to each unit if desired. Depending on your instructional goals, you may take 2 or 3 days to explore the mathematicians, ask students to participate in discussions, or respond to questions you pose for essays. When introducing each new mathematician, you might choose to include that individual's picture on a class bulletin board (see Introductory Lesson) or on the world map (see Unit 1). At the conclusion of your instruction, students will see the diversity of people working in the field and be better able to see themselves as mathematicians.

Introduction

THINKING LIKE SERIES

This book is one in a series, developed in conjunction with the Center for Gifted Education at William & Mary, intended to develop process skills in various content areas and enhance discipline-specific thinking and habits of mind through hands-on activities. Each book in the series focuses on a specific discipline and grade level:

- In *Thinking Like a Geographer*, students in grade 2 develop and practice geography skills, such as reading and creating maps, graphs, and charts; examine primary and secondary sources; and think spatially on a variety of scales
- In *Thinking Like a Mathematician*, students in grade 3 engage in exploration activities, complete mathematical challenges, and then apply what they have learned by making real-world connections.
- In *Thinking Like an Engineer*, students in grade 4 complete design challenges, visit with an engineer, and investigate real-world problems to plan feasible engineering solutions.
- In *Thinking Like a Scientist*, students in grade 5 use inquiry-based investigations to explore what scientists do, engage in critical thinking, learn about scientific tools and research, and examine careers in scientific fields.

INTRODUCTORY LESSON
WHAT IS A MATHEMATICIAN?

What is a mathematician? What do mathematicians look like? What do they do every day? Where do they work? These are all valuable questions to pose to your students to understand their background and misconceptions.

As you begin your exploration of math concepts with high-ability students, you may choose to organize their thinking using a Frayer model. This organizer guides students' learning as they explore a new topic. In this case, students will explore the concept, "What is a mathematician?" Students may visualize many different people when they think of "mathematician." This initial conversation can act as a preassessment for future units. Listen carefully and be ready to dispel myths or misunderstandings.

What Is a Mathematician?

RESOURCES AND MATERIALS

- Introductory Lesson Frayer Model
- Large class Frayer model
- Student math notebooks
- Bulletin board with pictures and identifying information of modern mathematicians (assemble in advance; consider using a diverse range of people, such as Julia Robinson, Srinivasa Ramanujan, Katherine Johnson, and Dorothy Vaughn, or local mathematicians from colleges and universities in your area)
- Photo booth with math tools/props (e.g., rulers, tape measures, protractors, compasses, an abacus, thermometers, calculators, set squares, slide rules, scales, pan balances, a telescope, lab tools, and Galileo's thermometer)

ESTIMATED TIME

45 minutes

OBJECTIVES

In this lesson, students will:
- use a Frayer model to brainstorm characteristics of a mathematician,
- discuss what attributes make for a successful mathematician, and
- begin to envision themselves as mathematicians.

PRIOR KNOWLEDGE

Students will need to know how to share ideas thoughtfully and politely, especially when disagreeing with peers.

INSTRUCTIONAL SEQUENCE

1. Distribute Introductory Lesson Frayer Model and pose the following question: *What is a mathematician?* Give students time to complete the handout and discuss with a partner.
2. Record students' suggestions on a large class Frayer model. As students make suggestions, there may be disagreement on whether the ideas make sense. For the purposes of brainstorming, record each of the ideas.
3. Review each quadrant in the Frayer model, posing questions to guide students' discussion. Consider questions like:
   - What do mathematicians do?
   - How do they act?
   - What are their personalities like?
   - What jobs/hobbies do they have? What jobs/hobbies do they not have?
   - Is there an opposite of a mathematician?

THINKING LIKE A MATHEMATICIAN

4. At the end of the conversation, ask students to choose the best 5–7 ideas in each quadrant of the Frayer model. Ask: *What makes these ideas the best? How do they help us understand what a mathematician is?*
5. Present the mathematician bulletin board to your students (see Resources and Materials). This board can be used throughout the following units, with students adding information about other mathematicians as they research.
6. Point out the diversity represented on the mathematician bulletin board before announcing that there are some mathematicians very close by whom you would like to add . . . the students! Set up a photo booth with math tools and props to take a photo of each student to add to the board.
7. Discuss students' math notebooks. Students may select pictures of mathematicians (and themselves) for the notebook cover. Explain that they will build their math notebooks to use as a reference throughout the year.

EXTENSION ACTIVITIES

- Have students write a letter to a mathematician in the community posing some of the questions from the discussion.
- Have students create a chart connecting the character traits of mathematicians to books that they have read.

ASSESSMENT OBSERVATION

Students should respond to the question "What is a mathematician?" in their math notebooks.

NAME: _____ DATE: _____

INTRODUCTORY LESSON
Frayer Model

Directions: Complete the model, listing characteristics, examples, and nonexamples of a mathematician. Then, provide a working definition of "mathematician."

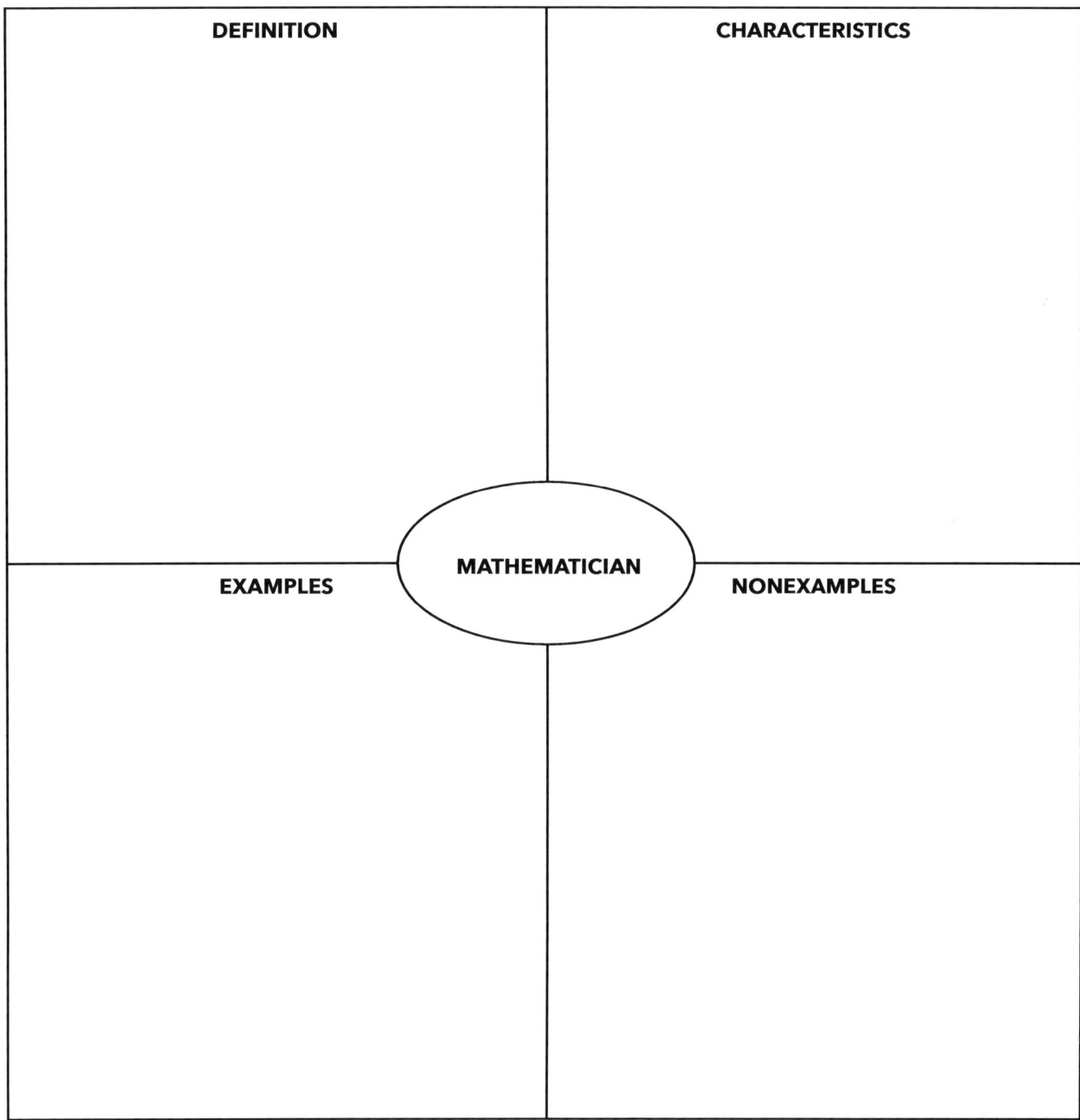

Thinking Like a Mathematician © Prufrock Press Inc.

UNIT 1
SYSTEMS OF COUNTING

RATIONALE

This unit shows students different ways of counting. Students will learn that the base-10 numbering system is not always used. The binary system will be introduced, and students will learn that computers use the binary system. This will solidify students' understanding of place value and how the number system works. This unit will also introduce basic computer concepts to students.

PLAN

In Lesson 1.1, students will explore the base-10 numbering system. Students will also explore other base numbering systems. In Lesson 1.2, students will apply different numbering systems to write binary code. In Lesson 1.3, students will discover different ways of counting around the world as they develop and present a project.

LESSON 1.1
EXPLORE DIFFERENT COUNTING SYSTEMS

RESOURCES AND MATERIALS

- Lesson 1.1 Base-4 Counting
- Three stackable cups with the numerals 0, 1, 2, and 3 written around each cup lip (per group; see Figure 1 for reference)
- Student math notebooks

ESTIMATED TIME

40–45 minutes

OBJECTIVES

In this lesson, students will:
- discover a pattern from a base-4 numbering system,
- understand the differences between numbering systems, and
- compare and contrast the base-10 numbering system with a base-4 numbering system.

PRIOR KNOWLEDGE

Students will need a basic understanding of a base-10 number system. They should also understand place value and how to add and subtract three-digit numbers.

Unit 1: Systems of Counting

INSTRUCTIONAL SEQUENCE

Anticipatory Set (5–10 minutes): Have students discuss the following in pairs: *Explain how our number system works. When you count, is there a pattern? What is the pattern? What would happen if there were no numerals after the numeral 3? How would that kind of counting work?*

Activity (30 minutes):
1. Divide students into groups of 2–3. Each group should receive three cups. Distribute Lesson 1.1 Base-4 Counting.
2. Have students stack the cups and turn them sideways so the numerals are showing (see Figure 1). Students should start with the number 000. Explain that (moving from left to right) cup 1 represents the hundreds digit, cup 2 represents the tens digit, and cup 3 represents the ones digit. Students will be turning the cups to find a pattern and recording their numbers on Lesson 1.1 Base-4 Counting.

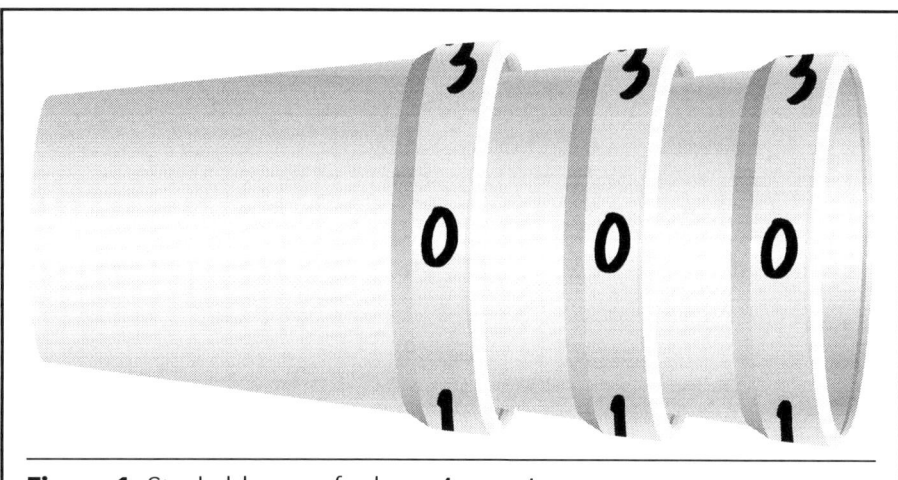

Figure 1. Stackable cups for base-4 counting.

3. Have students turn the ones digit (cup 3) to 1 and record the number shown: 001. They should then turn the ones digit again, and write the corresponding numbers until they reach the number 003.
4. Once students reach the number 003, have them turn the ones digit back to 0, and then turn the tens digit (cup 2) to 1. They should write down the number: 010.
5. Students should continue the pattern until they have reached 333.
6. Explain that this number system is called a base-4 number system (consisting of four numerals: 0, 1, 2, 3).
7. Lead a whole-group discussion about patterns of numbers.

Exit Ticket (5 minutes): Have students respond to the following in their math notebooks: *How would our numbering system change if there were no numeral zero? What if there were only the numerals 0 and 1?*

THINKING LIKE A MATHEMATICIAN

EXTENSION ACTIVITY

Have students add and subtract the following numbers in base 4, using carrying and borrowing rules. Students may also create their own addition and substraction problems.
- 110 + 011
- 223 + 121
- 323 − 122
- 301 − 223

ASSESSMENT OBSERVATIONS

- Students should discuss patterns in the base-4 numbering system.
- Students should be using the correct counting process with their cups: After students reach 003, they should reset cup 3 to the digit 0 and turn cup 2 to the digit 1.
- Students should discuss place value similarities and numeral differences.

NAME: _____ DATE: _____

LESSON 1.1
Base-4 Counting

Directions: Fill in the chart with the base-4 digits as you count the numbers. Then answer the questions on the next page.

000	001		
010			

Base-4 Counting, continued

1. Write the patterns you see in the new number system.

2. How is this counting system similar to the way we count numbers? Explain.

3. How is this counting system different from the way we count numbers? Explain.

4. What happens when you get to the highest possible ones digit?

5. What happens when you get to the highest possible tens digit?

LESSON 1.2

APPLY BINARY COUNTING

RESOURCES AND MATERIALS

- Lesson 1.2 Robot Maze
- Lesson 1.2 Writing in Binary Code
- Student math notebooks
- Coins (one per student)

ESTIMATED TIME

60–65 minutes

OBJECTIVES

In this lesson, students will:
- create a written block code to guide a robot through a maze,
- learn how binary code is formed, and
- create and translate a message in binary code.

PRIOR KNOWLEDGE

Students should understand how the base-10 numbering system works. They should also understand place value and how to add and subtract three-digit numbers.

INSTRUCTIONAL SEQUENCE

Anticipatory Set (5 minutes): Distribute Lesson 1.2 Robot Maze. Have students trace the way from start to finish, using only straight line segments.

THINKING LIKE A MATHEMATICIAN

Activity (45–50 minutes):

1. Place students in groups of 2–4. Explain that students will be programming a "robot" to move through the maze. Moving one square forward is equal to one unit. The robot begins in the "Start" square, facing north (top of the page).
2. Students should work together to create directions for the robot.
3. Tell students: Remember, the robot can only do exactly what you tell it to do. If you tell the robot to go forward, you must tell it how far to go, such as one unit. If you want to tell the robot to turn, you must include the direction and rotation, such as 90 degrees (or a quarter turn) clockwise.
4. Project the maze on the board.
5. Have students read their directions aloud, and draw a line where each student directs the robot. If a student does not specify a distance for movement, continue drawing a straight line until you reach the end of the board. Continue following students' instructions until you successfully complete the maze.
6. Explain to students that our number system is a base-10 system. Ask students to think about using a numbering system other than base 10. Have them brainstorm other base systems that we use (e.g., base 60 or base 12 to tell time on a clock).
7. Explain to students that computers are only able to understand binary code, or a base-2 system. All computer codes are turned into a string of binary numbers.
8. Distribute Lesson 1.2 Writing in Binary Code. Have students write the binary numbers 0000 to 1010 on the recording sheet. Explain to students that they will be using the same process they used in the base-4 system in Lesson 1.1.
9. Explain how letters are created in binary code. Each uppercase letter is represented as a string of eight digits beginning with 01, and each lowercase letter is represented as a string of eight digits beginning with 011. A is 01000001, B is 01000010, C is 01000011, D is 01000100, etc.; a is 01100001, b is 01100010, c is 01100011, d is 01100100, etc.
10. Instruct students to find the pattern for binary letters. Some students may see a pattern from these few letters. Some students may need to continue to write the alphabet. Students should write the rest of the letters in binary code in their math notebooks. Students should then write their name in binary code.
11. Have each group encode a message in binary for another group to decipher. Groups should exchange papers, and students should decipher each other's messages.

Exit Ticket (10 minutes): Have students flip a coin five times, recording a 1 for heads and a 0 for tails in order from left to right in their math notebooks. They

Unit 1: Systems of Counting

should then translate the resulting binary code into a number in our base-10 numbering system, including an explanation of how they translated their number.

EXTENSION ACTIVITY

Have students complete the following prompts:
- Add the following binary numbers: $010 + 001$, $011 + 010$, $101 + 111$.
- Subtract the following binary numbers: $011 - 001$, $010 - 001$, $100 - 001$, $1010 - 0101$.

ASSESSMENT OBSERVATIONS

- Students should write several successive digits in binary.
- Students' secret messages should indicate understanding of binary code.
- Students should explain how they translated their binary code in the Exit Ticket.

Lesson 1.2
Robot Maze

Directions: Using straight lines, create a path for the robot to take to reach the end of the maze. Then write your robot program code. Remember: When directing the robot, you must include the number of units (e.g., "move forward one unit"), as well as the direction and rotation of turns (e.g., "turn clockwise 90 degrees").

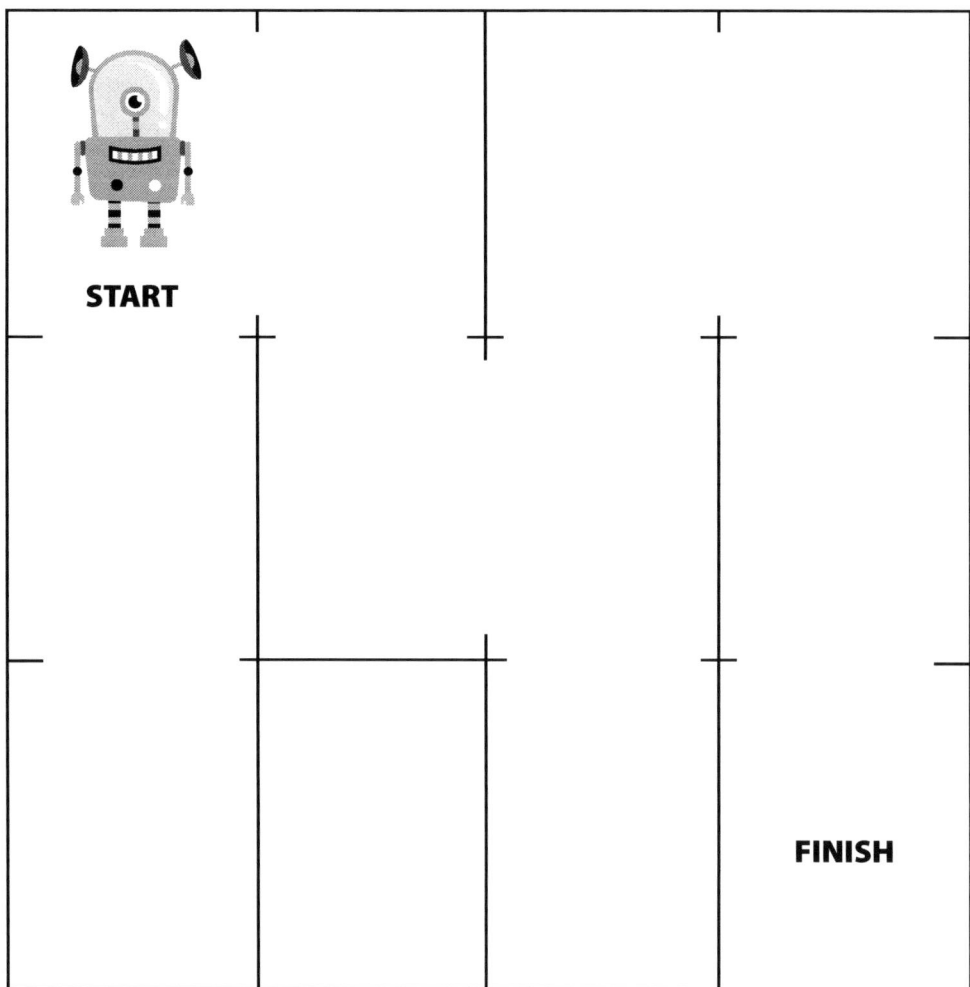

CODE:

NAME: _____ DATE: _____

LESSON 1.2
Writing in Binary Code

Directions: Answer the following prompts about binary code. Then, exchange papers with another group and decode that group's message.

1. Record the binary numbers 0000 through 1010.

2. Write your name in binary code.

3. Write a coded message to another group in binary code.

4. Exchange papers and decode your classmates' message.

Thinking Like a Mathematician © Prufrock Press Inc.

LESSON 1.3

PROJECT: SYSTEMS OF COUNTING AROUND THE WORLD

RESOURCES AND MATERIALS

- Lesson 1.3 Counting Project
- Lesson 1.3 Counting Presentation Page
- Lesson 1.3 Reflection
- Lesson 1.3 Rubric
- Large world map (posted on the class bulletin board or wall)

ESTIMATED TIME

Will vary depending on class interest and needs; at minimum 3 days, with 45 minutes per day

OBJECTIVES

In this lesson, students will:
- explore a system of counting,
- create a world map showing the location of the system's use, and
- give an oral presentation about the system of counting.

PRIOR KNOWLEDGE

Students should have an understanding of the base-10 numerical system and basic knowledge of world geography.

INSTRUCTIONAL SEQUENCE

1. Brainstorm essential questions about systems of counting with your students. Have them consider ideas relating to who, what, when, where, why, and how. Make a list of their questions on the board so that they can visually reference them for inspiration during their research.

Unit 1: Systems of Counting

2. Introduce the project. Tell students that they will research a system of counting and become experts. Distribute Lesson 1.3 Counting Project for students to begin their research.
3. Have students complete Lesson 1.3 Counting Presentation Page. Students should include the symbols of the numbers from 1–10.
4. When the students have completed their work, they should present their research (see Lesson 1.3 Rubric for presentation guidelines). You can then pin their work to the corresponding part of the class world map.
5. Throughout the project, have students reflect on their progress using Lesson 1.3 Reflection.

TEACHER NOTES

In this lesson, students will explore a system of counting and then share their discoveries by creating a world map bulletin board. There are numerous options for students to explore. You may ask them to work independently or in small groups in order to complete their research and presentation. Here are several options for students to explore:

- Sumerian counting
- Cuneiform numbers
- Mokshan numerals
- Babylonian numbers
- Aegean numerals
- Quipu
- Binary system
- Roman numerals
- Arabic numerals
- Decimal system, base 10
- Duodecimal system, base 12
- Vigesimal notation, base 20
- Daiji, used in Japan
- Hebrew alphabetic numerals

Research can be difficult for young students. As they work, they may be tempted to copy words directly from their resources. Guide their progress by discussing the importance of paraphrasing or summarizing other authors' writing. Plagiarism carries serious consequences for adults, and this is an opportunity for students to improve their skills.

Reassure students that they will have to explore multiple resources to find answers to their questions. Work with your librarian to develop a list of online and text resources for students to explore. Third graders should work online with direct adult supervision. They will have questions about this complex research project, and they will need guidance to stay on child-friendly websites.

THINKING LIKE A MATHEMATICIAN

EXTENSION ACTIVITIES

- Add examples of numerals to the class world map. The numerals should correspond to different number systems around the world. Ask students to notice similarities and differences among geographically close civilizations.
- Challenge students to develop a new system of counting using symbols. They should investigate problems and determine solutions for the new method.

NAME: _____ DATE: _____

LESSON 1.3
Counting Project

Directions: Research your system of counting and become an expert! Explore several different resources. You will probably not find all of the answers in one place. Remember, research is like a treasure hunt.

Your counting system: _____

History

1. Where is your system used? Find a map and locate where your system is used.

2. What language do people speak in that region of the world *today*?

3. When was the system of counting used?

4. What language did people speak *then*?

5. How did the system of counting develop?

NAME: _____ DATE: _____

Counting Project, continued

6. Who is a famous person who used your system of counting?

7. What years did he or she live?

8. How did this person use the system of counting?

How to Use the System

1. Learn to count from 1–10 in your system. Represent the numbers in symbols and words. Then explain how each number is pronounced.

What does the symbol look like?		How do you pronounce the number?
In the United States	**In your system**	
0		
1		
2		
3		
4		
5		
6		
7		
8		
9		
10		

NAME: _____ DATE: _____

Counting Project, continued

2. How do you represent the number 100 in your system? Explain how to create this number in step-by-step directions.

   Step 1:

   Step 2:

   Step 3:

3. What are some examples of where your system is used?

4. Explore your own questions. Find detailed answers.
   Question 1:

   Answer:

   Question 2:

   Answer:

NAME: _____ DATE: _____

LESSON 1.3
Counting Presentation Page

Directions: In preparation for your presentation, practice finding your region on a world map. Then, write the numbers 0–10 to show your classmates how to count.

Your counting system: _____

From the country/region of: _____

1. Draw the symbols of the numbers, counting from 1–10:

1		2	
3		4	
5		6	
7		8	
9		10	

2. Write a well-organized paragraph about your system of counting. Use a separate sheet of paper if necessary.

Thinking Like a Mathematician © Prufrock Press Inc.

NAME: _____ DATE: _____

LESSON 1.3
Reflection

Directions: Use this page to help you reflect on your project as you create it. You will reflect before, during, and after you create the project.

Your teammates (if applicable):

Dates that you reflected with your teacher: _____ _____ _____

Dates that you reflected with your team: _____ _____ _____

Reflect Before You Begin

1. What do you know about your topic?

2. What are your goals for this project?
   a. I want to learn more about:

   b. I want to get better at:

3. What resources do you need?

Thinking Like a Mathematician © Prufrock Press Inc.

NAME: _____ DATE: _____

Reflection, continued

4. Do you need help finding anything? If so, ask your teacher to suggest resources in the space below.

Reflect as You Work

1. What are you most proud of in your project so far?

2. What are you doing currently to work well with your team?

3. What is not going well with your project?

4. What aspects of the project would you like more help to accomplish? Ask your teacher if you need any guidance or resources.

5. What strategies can you use to improve?

NAME: _____ DATE: _____

Reflection, continued

6. What changes will you make to improve your project?

Reflect on Your Finished Project

1. Overall, what do you think of your finished project?

2. Why do you think your project turned out this way?

3. What is the best part of your project?

4. What part of your project could you improve?

NAME: _____   DATE: _____

LESSON 1.3 RUBRIC
Counting Project

Criteria	Points Possible	Points Earned
Research		
Student used research time wisely, making appropriate progress daily.	40	
Student found a detailed level of essential information.	40	
Student's research was well organized and neat.	20	
Total:	100	
Presentation		
Student presented project professionally, standing with tall posture.	25	
Student explained essential aspects of research with detail: • Where the system is used. • The symbols and their meanings. • History of the system.	50	
Student spoke loudly and clearly.	25	
Total:	100	
Work Ethic		
Student's work was completed and submitted on time.	50	
Student's work was neat, tidy, and presentable.	50	
Total:	100	

UNIT 2
ORDER OF OPERATIONS

RATIONALE

High-ability students in grade 3 may be ready for problems with multiple operations. Direct instruction will be essential for their success, as many of their past experiences probably relied mostly on one operation. Within this unit, students might encounter unknown mathematical calculations. Project-based learning is designed to present students with questions and then embed instruction within teachable moments. You might find that you need to introduce the concept of multiplying with decimal points or discuss multiplication as repeated addition. Students may also need to work through and revisit the calculations in order to formulate thoughtful questions. Your role in this process is to recognize your students' needs and facilitate meaningful learning.

PLAN

In Lesson 2.1, students will complete an exploration activity about the order of operations in addition, subtraction, multiplication, and division. In Lesson 2.2, students will complete an application activity about the order of operations with parentheses and exponents. In Lesson 2.3, students will complete a project involving the order of operations.

LESSON 2.1

EXPLORE THE ORDER OF OPERATIONS

RESOURCES AND MATERIALS

- Lesson 2.1 Calculating Order of Operations
- Student math notebooks
- Four-function calculator that does not perform order of operations (per group)
- Graphing/scientific calculator (per group)

ESTIMATED TIME

50–65 minutes

OBJECTIVES

In this lesson, students will:
- apply the order of operations with addition, subtraction, multiplication, and division; and
- evaluate problems when the order of operations is used incorrectly.

PRIOR KNOWLEDGE

Students should have an understanding of addition, subtraction, multiplication, and division of single-digit numbers. They should also have some experience using a calculator and finding differences that are less than zero.

INSTRUCTIONAL SEQUENCE

Anticipatory Set (10–15 minutes):
1. Ask students to brainstorm activities that require an order. As a whole class, discuss when things must occur in order (e.g., cooking, getting dressed, etc.).

Unit 2: Order of Operations

2. Present students with the following problem: *How would you simplify 3 × 7 − 2 + 4 ÷ 2?* Have them work alone in their math notebooks to reach a solution.
3. Then, instruct students to explain to a neighbor how they simplified the equation. What did they know that helped them simplify this question?
4. Discuss as a class: *Did you and your neighbor have the same answer? Why or why not?*

Activity (35–40 minutes):
1. Divide students into groups of 2–3. Half of the groups should have a four-function calculator, and the other half should have a graphing calculator.
2. Distribute Lesson 2.1 Calculating Order of Operations. Using their calculators, have groups simplify the expression and record their results.
3. Ask the groups to compare and discuss the results they got using different calculators. Ask: *Why would the answers be different? Can both of these answers be correct?*
4. Explain that when solving an expression, the order matters. Multiplication and division occur first from left to right. Then, addition and subtraction occur from left to right.
5. Tell students to simplify the question again without a calculator, using order of operations rules: 3 × 7 − 2 + 4 ÷ 2.

Exit Ticket (5–10 minutes): Have students simplify the following expressions in their math notebooks:
1. 3 + 2 × 5 − 10 ÷ 2
2. 12 ÷ 4 − 6 × 1 + 10 − 3
3. 24 − 6 ÷ 2 × 3 − 10 + 5 ÷ 5
4. 5 − 5 + 2 ÷ 1 × 3

EXTENSION ACTIVITY

Students who already understand order of operations with single-digit numbers may use algebra tiles, square pieces of paper, or graph paper to express an order of operations problem visually. Students will then discuss the different solutions if the order of operations is not used correctly.

ASSESSMENT OBSERVATIONS

- Students should be able to complete an order of operations problem in the correct order.
- Students should be able to describe the difference in using the two types of calculators.

NAME: _____ DATE: _____

LESSON 2.1

Calculating Order of Operations

Directions: Using your group's calculator, answer the following questions.

1. Are you using a four-function calculator or a graphing calculator?

2. Using your group's calculator, simplify: $3 \times 7 - 2 + 4 \div 2$

3. Compare your result to the result of a group that used a different type of calculator. Are the answers the same?

4. Discuss why your answers are the same or are different. Why do you think this happened?

36   *Thinking Like a Mathematician* © Prufrock Press Inc.

NAME: _____ DATE: _____

Calculating Order of Operations, continued

5. Do you think there are multiple correct answers? Why or why not?

6. Without a calculator, and using order of operations, simplify the expression: $3 \times 7 - 2 + 4 \div 2$

7. Which calculator gave this answer originally?

8. Explain why this might be.

LESSON 2.2

APPLY THE ORDER OF OPERATIONS

RESOURCES AND MATERIALS

- Lesson 2.2 Parentheses in Order of Operations
- Lesson 2.2 Exponents in Order of Operations
- Lesson 2.2 Order of Operations Guide
- Graphing calculator (per group)
- Student computers or tablets with Internet access

ESTIMATED TIME

55–60 minutes

OBJECTIVES

In this lesson, students will:
- discover the order of operations with parentheses and exponents, and
- apply all order of operations rules to multistep problems.

PRIOR KNOWLEDGE

Students should understand addition, subtraction, multiplication, division, and squaring numbers.

INSTRUCTIONAL SEQUENCE

Anticipatory Set (5 minutes): Ask students if they know any other operations (other than addition, subtraction, multiplication, and division) that might be used to simplify an expression. Students should brainstorm operations.

Unit 2: Order of Operations

Activity (45 minutes):
1. Divide the class into an even number of groups and give each group a number. Also give each student in each group a letter (e.g., A–C or A–D; you will group students using these letters later on).
2. *The odd-numbered groups* should research parentheses in order of operations. Students will use graphing calculators to simplify the expressions on Lesson 2.2 Parentheses in Order of Operations. They should then complete the rest of the problems.
3. *The even-numbered groups* will research exponents in order of operations. Students will use graphing calculators to simplify the expressions on Lesson 2.2 Exponents in Order of Operations. They should then complete the rest of the problems.
4. Once students have completed the problems, group students based on their letters (all A's together, all B's together, etc.). Students should explain their research and discoveries about parentheses or exponents to their new group members.
5. Direct students to complete the lesson handout that they had not previously completed (Parentheses or Exponents), using the information they learned from their letter groups.
6. Lead a class discussion about the order of operations, PEMDAS: Parentheses, Exponents, Multiplication, and Division (from left to right), Addition and Subtraction (from left to right). Distribute Lesson 2.2 Order of Operations Guide for reference.

Exit Ticket (5–10 minutes): Students should simplify the following expressions, using the order of operations.
1. $(3 + 1)^2 - 4 \times 2 + (8 - 7) + 3^2$
2. $12 \times 2 - (10 + 1) + 6^2 \div 3 + (12 - 10)$
3. $10 \div 2 - (2 - 1)^2 + 5^2 \times 1 \div 5$

EXTENSION ACTIVITIES

- Have students research order of operations with higher order exponents.
- Have students simplify multiple mathematical expressions involving higher order exponents and several sets of parentheses.
- Ask students to simplify the following expressions:
    1. $2^3 + 4 \times 2 \div 1 - 1^4$
    2. $(4 - 2)^4 - 12 \div 4 \times 2 + 3^3$
    3. $100 - 4^3 + (10 - 8)^2 + 3 \times 6 \div 3^2$

ASSESSMENT OBSERVATIONS

- Students should be able to complete an order of operations problem in the correct order.
- Students should understand how parentheses and exponents affect the order of operations.

NAME: _____   DATE: _____

LESSON 2.2

Parentheses in Order of Operations

Directions: Complete the following problems about parentheses and the order of operations.

1. Using a graphing calculator, simplify: $(3 + 2) \times 5 - 12 + (6 \times 3)$

2. How do the parentheses change the order of operations?

3. Using available resources, research how parentheses are used in the order of operations. Record your research here.

4. Simplify (no calculator allowed):
    a. $3 - (2 + 1) + 12 \div (2 \times 2)$

    b. $(4 \times 5) + (10 - 2) - 8 \div 4$

5. Create your own expression, and simplify your expression (no calculator allowed).

NAME: _____ DATE: _____

LESSON 2.2

Exponents in Order of Operations

Directions: Complete the following problems about exponents and the order of operations.

1. Using a graphing calculator, simplify: $3 + 2^2 \times 5 - 12 + 3^2$

2. How do the exponents change the order of operations?

3. Using available resources, research how exponents are used in order of operations. Record your research here.

4. Simplify (no calculator allowed):
   a. $25 + 3^2 - 4^2 \times 2$

   b. $30 + 5^2 + 6 \times 2^2 - 10$

5. Create your own expression and simplify your expression (no calculator allowed).

NAME: _____ DATE: _____

LESSON 2.2
Order of Operations Guide

Mathematicians solve problems in a specific order, following rules called the Order of Operations. The problems can be complex. Mathematicians always show every step so that they can check their work or answer questions if other people ask.

Step 1: Parentheses

Parentheses are curved marks that enclose a problem or expression, like this: (4 + 2)
Solve any problem on the inside of the parentheses first.

Step 2: Exponents

An exponent indicates the number of times to multiply the base number. For example, 2^3 means $2 \times 2 \times 2 = 8$
The exponent is not the same as $2 \times 3 = 6$.

Step 3: Multiplication and Division

Next, look for any multiplication and division. Then, work from left to right to solve.

Step 4: Addition and Subtraction

Lastly, solve any addition and subtraction problems. Work from left to right to solve.

Example

Look at this example. Notice how every step of the work is shown.

Problem	$2 \times 3 + (9 + 1) - 2^3$
Step 1: Parentheses	$2 \times 3 + 10 - 2^3$
Step 2: Exponent	$2 \times 3 + 10 - 8$
Step 3: Multiplication/Division	$6 + 10 - 8$
Step 4: Addition/Subtraction	$16 - 8$
Solution	8

LESSON 2.3

PROJECT: REFLECT ON THE ORDER OF OPERATIONS

RESOURCES AND MATERIALS

- Lesson 1.3 Reflection
- Lesson 2.3 Equation Cards (cut out in advance)
- Lesson 2.3 Order of Operations Project
- Lesson 2.3 Rubric

ESTIMATED TIME

Three 45-minute sessions; can vary

OBJECTIVES

In this lesson, students will:
- build upon their knowledge of the order of operations, and
- create a product explaining the order of operations.

PRIOR KNOWLEDGE

Students should understand basic addition, subtraction, multiplication, and division, as well as the signs for each operation.

INSTRUCTIONAL SEQUENCE

1. Introduce the project. Students will explore the question, "Does the order in which we calculate matter in math?" Teams work together on the calculations before creating a product to teach others the importance of the order of operations.

THINKING LIKE A MATHEMATICIAN

2. Give each team one equation card (see Lesson 2.3 Equation Cards) and a copy of Lesson 2.3 Order of Operations Project.
3. This project asks students to not only solve the equation correctly using the order of operations, but also develop wrong answers that reflect a common mistake a mathematician might make. *Note.* It is essential that you model how to solve the equation three ways and reassure students that is acceptable to develop two wrong answers. The order of operations also requires students to show their work. Advanced students may try to do the work in their heads, but this project requires them to show their work for each step of the problem.
4. Check students' work, either in small groups or individually. This could be a great time to reflect with the students on their teamwork and help resolve any issues (see Lesson 1.3 Reflection).
5. Ask teams to share their work in a way that best fits your classroom. Students may use Lesson 2.3 Rubric to guide them as they prepare their presentations.

Lesson 2.3
Equation Cards

$21 \div 3 + 2 - 4$	$12 + 6 \times 2 - 9$
$1 \times 3 + 10 \div 1$	$(6 + 4) \times (10 - 8)$
$(20 \div 2) - 3 + 5$	$(8 + 2) \times (3 + 7)$
$4 \times 10 - (2 + 9)$	$2 + 3^2 - 4 \times 2$
$(8 - 2) + 5^2 - 3$	$1 + 4 - 1 \times 3^2$

NAME: _____ DATE: _____

LESSON 2.3

Order of Operations Project

Directions: Answer the following questions about your group's equation card.

Our expression is:

Step 1: Simplify the expression correctly. Show each step.

Step 2: What could be a mistake another student could make? Show what that mistake looks like. Record each step.

Step 3: What could be a mistake another student could make? Show what that mistake looks like. Record each step.

NAME: _____ DATE: _____

Lesson 2.3
Rubric

Criteria	Points Possible	Points Earned
Project		
Student used work time wisely, making appropriate progress daily.	15	
Student's work was well organized and neat.	15	
Student represented the correct answer to the equation.	10	
Student identified two possible incorrect solutions.	20	
Student solved the equation to represent the possible mistakes.	20	
Student showed each step of work in all three solutions.	20	
Total:	100	
Presentation		
Student presented project professionally, standing with tall posture.	10	
Student accurately explained the correct solution to the equation.	20	
Student accurately explained two incorrect solutions.	20	
Student used math vocabulary during the presentation.	10	
Student referred to the order of operations rules throughout the presentation.	10	
Student spoke loudly and clearly.	10	
Student responded to audience questions thoughtfully.	10	
Student was a thoughtful audience member.	10	
Total:	100	

UNIT 3
PATTERNS

RATIONALE

Algebra is one of the five NCTM content standards. Students will discover patterns in numbers and in shapes, and apply these patterns to the real world. Pattern recognition is a major theme throughout algebra and geometry. This unit will allow students to recognize mathematics in the world around them by applying mathematical patterns to nature, art, and architecture.

PLAN

In Lesson 3.1, students will complete an exploration activity using sequences to explore the Fibonacci sequence. In Lesson 3.2, students will complete an application activity using the Golden Spiral and apply the Golden Spiral to nature and art. In Lesson 3.3, students will analyze patterns in famous works of art.

LESSON 3.1
EXPLORE PATTERNS

RESOURCES AND MATERIALS

- Lesson 3.1 Sequences
- Calculator (per group)
- Student computers or tablets with Internet access

ESTIMATED TIME

75–90 minutes

OBJECTIVES

In this lesson, students will:
- learn how to describe and predict numerical patterns algebraically,
- apply patterns to mathematical sets of numbers, and
- compare and contrast patterns of numbers.

PRIOR KNOWLEDGE

Students should understand:
- addition, subtraction, multiplication, and division;
- visual pattern recognition; and
- basic mathematical sequences of numbers.

Unit 3: Patterns

INSTRUCTIONAL SEQUENCE

Anticipatory Set (10 minutes): Have students complete the following two patterns.
1. 2, 4, 6, 8, ___, ___.
2.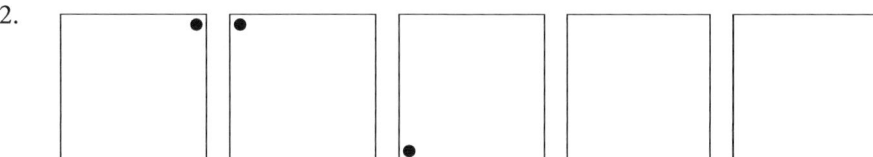

Activity (60–70 minutes):
1. Place students in groups of 2–4.
2. Explain that in mathematics a pattern is called a *sequence*. Display the sequence: 1, 4, 7, 10. Tell students to think about the sequence shown and record their observations on Lesson 3.1 Sequences.
3. Have students share their observations with their group members.
4. Work through the rest of the handout together, displaying the various sequences and allowing students to work on observations and explanations. Have students discuss their findings.
5. When students reach question 8, explain that this is the Fibonacci sequence. The Fibonacci sequence is a recursive sequence, found by adding the two prior terms. Lead a discussion about this sequence.

Exit Ticket (5–10 minutes): Have students reflect on the following prompt in their math notebooks: *What is your favorite pattern or sequence? Why? Create your own sequence.*

EXTENSION ACTIVITY

Ask students to find the nth term of the following sequences. These sequences begin with $n = 1$, and n increases by 1 each time. For example, the sequence 2, 4, 6, 8, . . . would be given by 2*n, because the sequence is equivalent to 2*1, 2*2, 2*3, 2*4, . . .

- If n begins at 1 and increases by 1 each time, write an expression for the nth term of the sequence 1, 4, 7, 10 . . .
- If n begins at 1 and increases by 1 each time, write an expression for the nth term of the sequence 1, 4, 9, 16, 25 . . .

ASSESSMENT OBSERVATIONS

- Students should be able to recognize patterns in shapes and in numbers (see Lesson 3.1 Sequences).
- Students should be able to describe the Fibonacci sequence.

NAME: _____ DATE: _____

LESSON 3.1

Sequences

Directions: Complete the following questions about the sequence 1, 4, 7, 10, . . .

1. Write your observations of the sequence 1, 4, 7, 10.

2. Explain how the terms in the sequence 1, 4, 7, 10 are found.

3. Write the next two numbers in the sequence 1, 4, 7, 10: _____, _____

4. Draw and explain a model that represents the sequence 1, 4, 7, 10, . . .
    a. Drawing:

    b. Explanation:

Thinking Like a Mathematician © Prufrock Press Inc.

NAME: _____ DATE: _____

Sequences, continued

5. Write your observations of the sequence 64, 32, 16, 8:

6. Explain how the terms of the sequence 64, 32, 16, 8 are found:

7. Write the next two numbers in the sequence 64, 32, 16, 8: _____ , _____

8. Write your observations of the sequence 1, 1, 2, 3, 5, 8, 13:

9. Write the next five terms of the Fibonacci sequence 1, 1, 2, 3, 5, 8, 13:

10. What do you notice about the numbers in the Fibonacci sequence?

11. What do you wonder about the Fibonacci sequence?

LESSON 3.2

APPLY PATTERNS TO THE GOLDEN SPIRAL

RESOURCES AND MATERIALS

- Lesson 3.2 The Golden Spiral
- Student math notebooks
- Graph paper (one sheet per student)
- Colored pencils/markers
- Student computers or tablets with Internet access

ESTIMATED TIME

55–65 minutes

OBJECTIVES

In this lesson, students will:
- create a graph of Fibonacci squares and the golden spiral,
- research the golden spiral in a field of their choosing, and
- reflect on math's relationship with other subject areas.

PRIOR KNOWLEDGE

Students should be familiar with the Fibonacci sequence (see Lesson 3.1) and how to use graph paper.

INSTRUCTIONAL SEQUENCE

Anticipatory Set (5 minutes): Have students write the first 10 terms of the Fibonacci sequence (see Lesson 3.1). Review together as a class.

Unit 3: Patterns

Activity (45–50 minutes):
1. For this activity, students may work in small groups or alone. Distribute graph paper and colored pencils or markers.
2. Ask students to color in one square at the center of the graph paper. Then, using a different color pencil, they should color a second square directly above the first square on the graph paper. These squares represent the first two numbers in the Fibonacci sequence (1, 1).
3. Using a different colored pencil, students should then color in a square with a side length of *two* boxes (representing the third number of the Fibonacci sequence) directly to the left of the two boxes already colored.
4. The students should then color in a square with side length of *three* boxes, directly below the other boxes; then a square with a side length of *five* boxes, directly to the right of the other boxes.
5. Encourage students to continue this pattern until the graph paper is full (see Figure 2).
6. Explain that all of the squares the students made have side lengths that correspond to the numbers in the Fibonacci sequence.

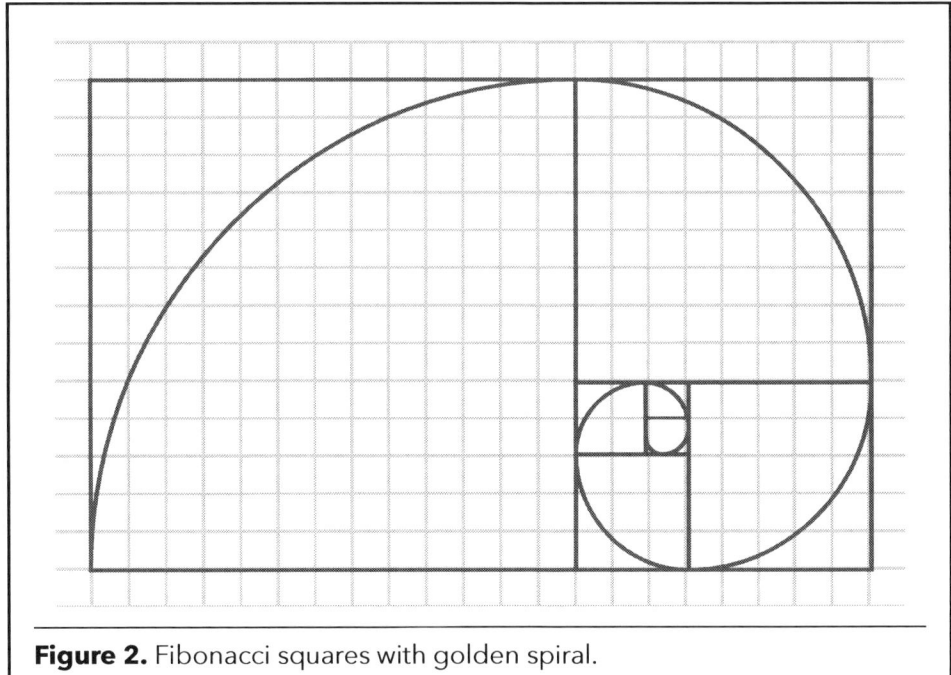

Figure 2. Fibonacci squares with golden spiral.

7. Once all of the boxes are colored, students should create a spiral as follows:
   - Start at the bottom left corner of the bottom square with a side length of one, and draw an arc to the top right corner of the same square with a dark marker or pen.
   - Continue the spiral to the top left corner of the next square with a side length of one.
   - Then spiral to the bottom left corner of the square with a side length of two.

THINKING LIKE A MATHEMATICIAN

- Connect this line to the bottom right corner of the square with side length of three.
- Continue this spiral by continuing in a counterclockwise direction, connecting opposite vertices of the squares (see Figure 2).

8. Ask students to share their work with each other.
9. Tell students that they have created a "golden spiral." Distribute Lesson 3.2 The Golden Spiral for students to complete.
10. Have students research the golden spiral in nature, using a computer.

Exit Ticket (5–10 minutes): Lead a class discussion: *Explain how math and art may be related. Discuss how you think a mathematician may apply math to a different subject.*

EXTENSION ACTIVITIES

- Have students divide successive terms in the Fibonacci sequence (e.g., 1/1, 2/1, 3/2, 5/3, etc.), using a calculator. Students should record at least 10 quotients, written as decimals, in their math notebooks. Explain that the numbers they have been finding are approaching "the golden ratio." Ask: *What number are these ratios approaching?*
- Have students research the golden ratio and its applications in a real-world field of the their choice (e.g., art, architecture, space exploration, etc.).

ASSESSMENT OBSERVATIONS

- Students should be able to create the golden spiral and find application in some other discipline. The golden spiral is found in nature in the shape of galaxies, in the way some plants grow, in the way some petals are formed, in the shape of some sea life, and in art.
- Students should apply the Fibonacci sequence and the golden ratio/golden spiral to a discipline of their choosing.

NAME: _____ DATE: _____

LESSON 3.2
The Golden Spiral

Directions: Consider the following questions about the golden spiral.

1. Explain how the squares you created on your graph paper correspond to the Fibonacci sequence.

2. How did you create the golden spiral?

3. Explain something you have seen before that resembles the golden spiral.

Extend Your Thinking

Type or write a paragraph summarizing your research about the golden spiral in nature. Be sure to include a works cited section.

LESSON 3.3
PROJECT: ANALYZE PATTERNS IN ART

RESOURCES AND MATERIALS

- Lesson 1.3 Reflection
- Lesson 3.3 Suggested Artists and Works of Art (teacher copy)
- Lesson 3.3 Art Patterns Project
- Lesson 3.3 Rubric
- Research materials, like websites or books about artists

ESTIMATED TIME

Will vary based on student engagement

OBJECTIVES

In this lesson, students will:
- research and analyze a famous work of art;
- discover artistic techniques, such as repeated use of color, line, and shapes; and
- create two pieces of their own art using the patterns they have discovered.

PRIOR KNOWLEDGE

Students should have knowledge of repeating, growing, and shrinking patterns in math, as well as the names of various shapes and colors.

Unit 3: Patterns

INSTRUCTIONAL SEQUENCE

1. Have students analyze a piece of artwork in a "think aloud" lesson. Point out the artist's repeated use of colors, lines, shapes, and textures. Your presentation will serve as a model for the students' research projects.
2. Help students choose a work of art to study. They may wish to browse books or websites to help them find a piece of art. See Lesson 3.3 Suggested Artists and Works of Art for examples; this handout may distributed to students or used for teacher reference.
3. Guide students to analyze their own work of art. Distribute Lesson 3.3 Art Patterns Project to aid them in their analysis.
4. Choose a work of art and model the project that students will be completing. Show students how they might create a new work of art, drawing inspiration from an existing piece. Use the same repeated element as you create the work.
5. Guide students to produce their own works of art using the same repeated element that the artist used.
6. Discuss with students how to present their artwork to the class. (See Teacher Notes.)

TEACHER NOTES

Students should hunt for repeated colors, lines, or shapes over the entire work of art. The artist may organize the patterns in a line across the work, in concentric or circular patterns, or spread throughout the art. For example, William Morris used repeated birds in several of his wallpapers. The birds are placed symmetrically, and he colored them with the same colors. The patterns, then, are repeated shapes and repeated colors.

Repeated colors. Artists like Henri Matisse relied on color to provide structure and interest to their work. Students can study Matisse's color choices in works like *The Parakeet and the Mermaid* (1952). In this work, Matisse repeated the same blue, green, orange, and plum colors across his approximately 11' x 25' collage.

Repeated lines. Many artists have to consider the layout of the image and, therefore, the lines in their work. Students may understand this concept best when they view black-and-white woodblock prints. Elizabeth Catlett created beautiful portraits using a variety of carving tools. *Sharecropper* (1970) uses numerous cuts to create patterns in lines, which in turn, sculpt a patterned sun hat, gradient clothing, and a woman's sculpted face.

Repeated textures. Artists use many different tools to create the look and feel for viewers. William Morris is a famous British artist known for developing patterns with his textiles. Artists like Ansel Adams, famous for his black-and-white photographs of nature, captured textures. You might consider developing a cross-curricular lesson with English to help students brainstorm and research texture vocabulary. As an introduction to this concept, have students consider words like *bulky, coarse, flawless, abrasive, dull,* or *ragged*.

THINKING LIKE A MATHEMATICIAN

Repeated shapes. Artists often choose to repeat shapes within their work. Andy Warhol is famous for using one image and manipulating the colors repeatedly to create new works of art. For example, he used one image of Marilyn Monroe and bright colors to create the silkscreen *Marilyn Monroe* in 1967.

Ask your students to explore these famous pieces of art (see Lesson 3.3 Suggested Artists and Works of Art for more). Some names and pieces are recognizable, while others are famous within the specific field of design. *Note.* Although this list of work is appropriate for third graders, some of these artists have represented nude models in their other works. If you choose to broaden the research component of this project, please guide students in the way that you feel most appropriate.

You and your students should also thoughtfully reflect on their work throughout the process, not just at the end (see Lesson 1.3 Reflection). Also included in this project is a rubric that allows for multiple grading opportunities.

NAME: _____ DATE: _____

Lesson 3.3
Suggested Artists and Works of Art

Numerous artists use repeated colors, lines, shapes, and textures. The works below are student-friendly pieces that exhibit patterns.

Artist	Name of the Work	Date	Best Element to Explore
Henri Matisse, painter and printmaker	*The Parakeet and the Mermaid*	1952	Repeated shapes, repeated colors
	The Swimming Pool (La Piscine)	1952	Repeated color
	The Clown (Le Clown)	1943	Repeated shapes, repeated colors
Jackson Pollock, painter	*Mural*	1943	Texture
	Full Fathom Five	1947	Repeated texture
Ansel Adams, photographer	*Mount Williamson–Clearing Storm*	1945	Repeated texture, shapes
	Dune, White Sands National Monument, New Mexico	1941	Repeated texture, colors
Frank Lloyd Wright, architect	*Fallingwater*	1935	Repeated texture, shapes
	Kentuck Knob	1954	Repeated shapes, lines
	Robie House	1906	Repeated shapes, lines
Alexander Calder, sculptor	Arc of Petals	1941	Repeated shape, texture
	Man	1967	Repeated lines, texture
	Untitled (Mobile at the National Gallery of Art, Washington, DC)	1976	Repeated shapes, line
Georges Seurat, painter	*A Sunday on La Grande Jatte*	1884	Repeated textures
	The Eiffel Tower	1889	Repeated textures
	The Circus	1890–1	Repeated texture
William Morris, textile designer	*Peacock and Dragon*	1878	Repeated shape, color
	Strawberry Thief	1883	Repeated shape

Thinking Like a Mathematician © Prufrock Press Inc.

NAME: _____ DATE: _____

LESSON 3.3
Art Patterns Project

Directions: You will be researching a famous work of art. Fill out the following information to describe and analyze your piece of art. Use a separate sheet of paper if necessary.

Name of the famous artist: _____

Name of the famous artwork you're studying: _____

When was it created? _____

Describe

What does your image illustrate?

What colors does the artist use?

What shapes does the artist employ?

When you first observe your art, what element repeats? (Circle one or more.)

        Colors   Shapes   Textures   Lines

Explain the repeating element, using as much math vocabulary as you can. For example, you might explain how the artist used repeating colors.

Choose a second repeating element. How did the artist repeat that element?

NAME: _____ DATE: _____

Art Patterns Project, continued

Analyze

What pattern stands out the most?

Why did the artist use a pattern?

How would the art be different if the pattern did not exist?

Create

Create two new pieces of art, inspired by the original:
- Artwork #1: Create a similar piece of work without a repeated pattern.
- Artwork #2: Change the pattern to create a new image. You must use the same elements that your artist did (repeated colors, shapes, textures, or lines), but repeat them in a different way.

Present

Plan a presentation of the three pieces of art (the piece you researched and the two pieces you created). You should first show your class the original artwork. Explain it to other students using your research. Then explain the two pieces of artwork that you created.

NAME: _____ DATE: _____

LESSON 3.3
Rubric

Criteria	Points Possible	Points Earned
Research Grade		
Student used research time wisely, making appropriate progress daily.	20	
Student found a detailed level of essential information.	25	
Student's written responses are thoughtful, detailed, and complete.	35	
Student's research is well organized and neat.	20	
Total:	100	
Written Product for Research/Analysis		
Student's written explanation is neat, and letters are properly formed.	10	
Student's writing includes proper grammar and spelling.	20	
Student used proper capitalization and punctuation.	20	
Student included the artist's name and the name of the work.	10	
Student included math vocabulary.	10	
Student explained the artwork in detail, including a thorough description and analysis.	30	
Total:	100	
Presentation Grade		
Student presented project professionally, standing with tall posture.	20	
Student spoke loudly and clearly.	20	
Student explained essential aspects of research with detail: • The artist's artwork. • Two student-created art pieces. • How the two new pieces are similar to the original art.	30	
Student displayed the artwork in a way that the audience could view.	10	
Student responded to audience questions with authority.	10	
Student was a thoughtful audience member.	10	
Total:	100	
Work Ethic and Reflection		
Student's work was completed and submitted on time.	25	
Student's work was neat, tidy, and presentable.	25	
Student thoughtfully reflected on work throughout the process.	25	
Student incorporated feedback in meaningful ways.	25	
Total:	100	

UNIT 4
TIME AND MEASUREMENT

RATIONALE

Students will discuss time and how days and years are relative to planetary movement. Students will learn about length of time on our planet, and length of time on other planets in our solar system. The activities and project included in this chapter will help students to grasp the relationship between planetary motion and time/seasons.

PLAN

In Lesson 4.1, students will complete an exploration activity calculating the lengths of days and years on other planets with respect to Earth's time. In Lesson 4.2, students will apply time measurements to seasons, and discuss how the change in a planet's elliptical path changes the length of the seasons. In Lesson 4.3, students will build models to explain the relationships between planets in the solar system.

LESSON 4.1
EXPLORE TIME ON PLANETS

RESOURCES AND MATERIALS

- Lesson 4.1 Planetary Research
- Student computers or tablets with Internet access and multimedia presentation capability

ESTIMATED TIME

90–120 minutes; will vary depending on amount of time given to research planets and complete the multimedia presentation

OBJECTIVES

In this lesson, students will:
- research day and year lengths on different planets,
- apply time measurement to other planets, and
- compare and contrast the length of daylight hours on Earth and other planets.

PRIOR KNOWLEDGE

Students should have an understanding of the length of a day, an hour, and a year on Earth. They should also understand the measurement of minutes in an hour, hours in a day, and days in a year.

Unit 4: Time and Measurement

INSTRUCTIONAL SEQUENCE

Anticipatory Set (10–15 minutes): Have students answer the following questions individually in their math notebooks.
1. How many hours are in our day?
2. How many minutes are in an hour?
3. How many minutes are in our day? Be sure to show your work.
4. What determines how many hours are in a day on Earth?
5. What determines how many days are in a year on Earth?
6. Do all planets have the same number of hours in a day and days in a year as Earth? Explain your reasoning.

Lead a class discussion about the questions, discussing how time is measured and how to select the correct measuring tool (e.g., *We do not want to measure temperature with a ruler, just like we do not want to measure a month in minutes or seconds*).

Activity (Time will vary):
1. Divide students into groups of 2–4 and distribute Lesson 4.1 Planetary Research. Tell students that they will be researching three other planets in our solar system, finding out the number of hours in a day and the number of days in a year on those planets (in Earth hours and days). Students should use the Internet to research. Have them keep track of what websites and sources they use.
2. Have students compare their research with another group. Ask: *What similarities do you see? What differences? Why might this be?*
3. Instruct student groups to create a multimedia presentation (e.g., PowerPoint, Sway, Flipgrid, Prezi, video, recorded song, etc.) on the three planets they researched.
4. Have students present the multimedia presentation to the class.

Exit Ticket (10 minutes): Tell students: *Based on what you have learned about the length of days and years on other planets, which planet would you want to live on? Write a brief paragraph about why you would want to live on this planet. (Your reasoning should be based on what you learned in this lesson.)*

EXTENSION ACTIVITIES

- Have students research and explain what changes the length of daylight hours on Earth.
- Have students research a second planet and determine if the length of daylight hours changes. If so, what determines the change of the length of daylight hours?

THINKING LIKE A MATHEMATICIAN

ASSESSMENT OBSERVATIONS

- Students should present a completed multimedia presentation to the class about time on planets.
- Students should identify the planet on which they want to live and justify their answers using information gathered in this unit.

NAME: _____ DATE: _____

LESSON 4.1

Planetary Research

Directions: Choose three planets you would like to research. Find out the number of hours in a day and the number of days in a year on these planets.

Where will you find information about your planets?

Planet #1: _____

How many Earth hours are in a day?

How many Earth days are in a year?

Planet #2: _____

How many Earth hours are in a day?

How many Earth days are in a year?

Planet #3: _____

How many Earth hours are in a day?

How many Earth days are in a year?

LESSON 4.2
APPLY SOLAR MATH

RESOURCES AND MATERIALS

- Lesson 4.2 Orbits of Planets
- Student math notebooks
- String (24 inches per group)
- Pencil (per group)
- Student computers or tablets with Internet access

ESTIMATED TIME

50–70 minutes

OBJECTIVES

In this lesson, students will:
- discover the properties of an ellipse, and
- apply elliptical orbits to the measurements of seasons on planets.

PRIOR KNOWLEDGE

Students should have an understanding of what a circle is, as well as Earth's seasonal changes.

INSTRUCTIONAL SEQUENCE

Anticipatory Set (5–10 minutes): Have students answer the following questions in their math notebooks:
1. What is your favorite season of the year? Why?
2. How are the seasons in our region of the world different than in another region?

Unit 4: Time and Measurement

3. Do you think there are seasons on other planets? What would cause the seasons on other planets?

Activity (45–60 minutes):
1. Place students in groups of two and give each group a pencil and a string.
2. Instruct Student A to hold both ends of the string still on a piece of paper at two points, so that the string is loose.
3. Student B should place a pencil somewhere on the string, and pull the string taught. Student B should then use the pencil to trace a shape, keeping the string taught, tracing around one side of Student A's fingers (see Figure 3).

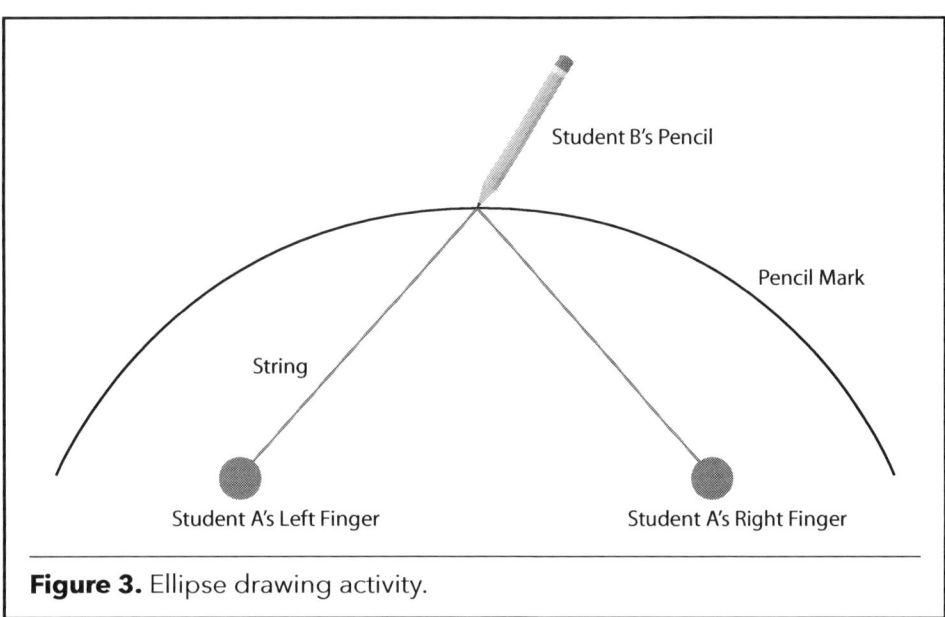

Figure 3. Ellipse drawing activity.

4. Tell the students to switch roles. Student B will hold the string in the same place, and Student A will draw the line on the opposite side.
5. Explain that the students just created a mathematical shape called an *ellipse*, and that the two points used for the ends of the string are called the *foci* (plural for *focus*) of the ellipse.
6. Explain how ellipses relate to planetary orbits: *According to Johannes Kepler, the planets orbit the sun in the shape of an ellipse with the sun as one of the foci.*
7. Distribute Lesson 4.2 Orbits of Planets. Tell students that they will be researching three planets, their orbital shapes, and how the orbital shapes determine the length of years and seasons. Students should use the Internet to research, keeping track of the websites they use.
8. (Optional) Students may add information to the multimedia presentation created in Lesson 4.1.

THINKING LIKE A MATHEMATICIAN

Exit Ticket: Have students answer the following prompt in their math notebooks: *Compare an ellipse and a circle. Write a short paragraph about how the ellipse and circle are similar and how they are different. How would a planet with a circular orbit be different than a planet with a very long elliptical orbit?*

EXTENSION ACTIVITIES

- Have students research the evolution of thought and discovery regarding the orbital motion of the planets, tracing the development from the ancient Greek philosophers to modern times.
- If students choose to add to their multimedia presentations from Lesson 4.1, have them present at a science fair or to another authentic audience.

ASSESSMENT OBSERVATIONS

- Students should present information about time and season measurements based on planets' elliptical orbits.
- Students should relate seasonal measurements to planetary orbit.

NAME: _____ DATE: _____

LESSON 4.2
Orbits of Planets

Directions: Select three planets to research. Use what you learn to answer the following questions about each planet. Use a separate sheet of paper or your math notebook to record your answers.

1. Is this planet's orbit more circular or more elongated?

2. What is the closest distance this planet comes to the sun?

3. What is the farthest distance this planet is away from the sun?

4. How long is a year on this planet?

5. How long are the seasons on this planet?

6. How does the shape of the orbit of this planet affect the length of a year and the seasons?

7. Are there any other factors that determine the seasons on this planet? If so, please explain.

Thinking Like a Mathematician © Prufrock Press Inc.

LESSON 4.3

PROJECT: DESIGN SOLAR MODELS

RESOURCES AND MATERIALS

- Lesson 1.3 Reflection
- Lesson 4.3 Days and Years on Planets
- Lesson 4.3 Planet Project
- Lesson 4.3 Rubric
- Materials for creating planet models (will vary based on students' needs)

ESTIMATED TIME

120 minutes for the lesson; time will vary for creating projects

OBJECTIVES

In this lesson, students will:
- explore the lengths of planets' days and years, and
- design and create a model to compare the lengths of days or years on two different planets.

PRIOR KNOWLEDGE

Students should understand that the length of days and years varies for each planet in our solar system.

Unit 4: Time and Measurement

INSTRUCTIONAL SEQUENCE

1. Introduce the research project, either assigning it to individuals or small groups.
2. Model how to research, including finding reliable resources and deciding how to record the information.
3. Distribute Lesson 4.3 Days and Years on Planets.
4. As a class, discuss different ways that students could represent their findings. You may choose to model an example or let the students use their own creativity. Distribute Lesson 4.3 Planet Project. Assign a deadline for their final projects.
5. Give the students time to complete their projects.
6. Allow students to present their projects to the class.

TEACHER NOTES

The math and science connection possibilities in this project are numerous, leaving plenty of opportunities to extend the explorations to fit students' interests. This project provides an opportunity for students to learn to collaborate. High-ability students often prefer to work independently, but it is important for them to learn to work effectively in a small group to share the workload during research. You might also assign parts of the research to small groups, with the expectation that each group will share the findings with the larger group.

As students complete their research, you may discover that different groups found slightly different data for the planets. Use this as a teaching opportunity to discuss verifying information using a variety of sources, the reliability of sources, and the scientific reason that the days might be measured differently (e.g., a solar day is measured slightly differently than a sidereal day).

The calculations section (Lesson 4.3 Days and Years on Planets) may be difficult for students, as it requires them to think flexibly about numbers. Because the equations do not work out perfectly, students may need you to model the strategies to solve these types of problems. You should discuss estimating in a variety of ways, and model multiple ways to solve problems. From there, allow the students to work with the estimation and solutions in order to help develop their number sense. You might allow them to check the work with calculators, but make sure the tool is only used to verify their answers.

As a culminating project, the students have flexibility to design and create a model to compare the length of days or years on two different planets. The project is purposefully open-ended, allowing students to direct their learning in ways that are meaningful to them. For example, the students could build a scale model of their planets to show the different sizes in orbital paths, or compose a musical piece to represent their findings. You may find that you want to give more direction, based on your group of learners.

NAME: _____ DATE: _____

Lesson 4.3
Days and Years on Planets

Directions: Research the movement of each of the planets. Record the amount of time that it takes for each planet to complete one rotation on its axis. Then find the amount of time it takes for each planet to complete one revolution around the sun.

Planet	Length of One Day	Length of One Year	Interesting Fact
Mercury			
Venus			
Earth			
Mars			
Jupiter			
Saturn			
Uranus			
Neptune			

NAME: _____ DATE: _____

LESSON 4.3
Planet Project

Directions: Build a model to compare two planets. Your model should represent the difference between the length of either the days or years. It should also explain why there is a difference.

The two planets I want to represent are:

1. _____

2. _____

   I want to show the length of (circle one):   days   years

The materials I need to create my models are:

1. _____
2. _____
3. _____
4. _____
5. _____
6. _____

To write my paragraph description, I plan to use these vocabulary words:

I need help from an adult to:

The due date for my project is _____

Thinking Like a Mathematician © Prufrock Press Inc.

77

NAME: _____ DATE: _____

LESSON 4.3
Rubric

Criteria	Points Possible	Points Earned
Research Grade		
Student used research time wisely, making appropriate progress daily.	20	
Student found a detailed level of essential information.	25	
Student's written responses are thoughtful, detailed, and complete.	35	
Student's research was well organized and neat.	20	
Total:	100	
Project		
Student's project accurately reflected the science and math concepts.	50	
Student's writing includes proper grammar and spelling.	15	
Student used proper capitalization and punctuation.	15	
Student included appropriate amounts of details in the project and writing.	20	
Total:	100	
Presentation Grade		
Student presented project professionally, standing with tall posture.	20	
Student spoke loudly and clearly.	20	
Student explained essential aspects of research with detail.	40	
Student responded to audience questions with authority.	10	
Student was a thoughtful audience member.	10	
Total:	100	
Work Ethic and Reflection		
Student's work was completed and submitted on time.	25	
Student's work was neat, tidy, and presentable.	25	
Student thoughtfully reflected on work throughout the process.	25	
Student incorporated feedback in meaningful ways.	25	
Total:	100	

UNIT 5
GRAPHING, DATA, AND CHARTS IN ALGEBRA

RATIONALE

This unit builds upon algebra, one of the NCTM content standards. Students should be able to analyze data given and then create a chart and a graph. This unit will allow students to compare and contrast linear versus nonlinear data. Students will discover patterns of linear and nonlinear data in a chart and a graph, and then will apply that data through a culminating project.

PLAN

In Lesson 5.1, students will complete an exploration activity on linear measurement. Students will evaluate how linear data correspond to a linear graph. In Lesson 5.2, students will complete an application activity on nonlinear data using hot or cold water, and temperature change over time. In Lesson 5.3, students will create their own business and build graphs to represent costs, income, and profit.

LESSON 5.1
EXPLORE LINEAR DATA, CHARTS, AND GRAPHS

RESOURCES AND MATERIALS

- Lesson 5.1 Graphing Time and Distance
- Student math notebooks
- Ruler (per group)

ESTIMATED TIME

35–40 minutes

OBJECTIVES

In this lesson, students will:
- interpret data (distance, time, and speed) in linear form,
- graph the data, and
- learn which graphs and charts help to explain and predict data.

PRIOR KNOWLEDGE

Students should have experience creating graphs and plotting points.

INSTRUCTIONAL SEQUENCE

Anticipatory Set (5–10 minutes): Have students respond to the following in their math notebooks: *How do we measure the speed a car travels? How do we measure distance? What are some measurements for time?*

Unit 5: Graphing, Data, and Charts in Algebra

Activity (25–30 minutes):
1. Divide students into groups of two and distribute Lesson 5.1 Graphing Time and Distance.
2. Students should consider the data in the chart and complete the questions together.
3. Discuss the questions as the students work, focusing on the mathematics of linear graphs. The y values will increase (or decrease) at the same rate for each equal increment of increase in x.

Exit Ticket (5 minutes): Have students consider the following in their math notebooks: *Explain how charts and graphs help us in the real world. Explain a situation in which a linear graph would help predict something.*

EXTENSION ACTIVITY

Have students complete the following questions using the information from Lesson 5.1 Graphing Time and Distance.
- How many miles per hour is Lindsay traveling?
- If Lindsay drives home from the store at double the speed she traveled to the store, how long will it take her to get home?
- If the speed limit is 45 miles per hour, is Lindsay speeding on her way to or from the store?

ASSESSMENT OBSERVATIONS

- Students should graph the data that correspond to the miles driven, and answer the discussion questions.
- Students should discover that the data is linear.
- When analyzing the numbers, students may answer that the numbers increase in equal amounts per consistent time interval. This will always be true for a linear graph. If the x values are given in equal increments, then the y values will increase or decrease in equal increments.

NAME: _____   DATE: _____

LESSON 5.1
Graphing Time and Distance

Directions: Lindsay travels from her house to the store, recording her time in the following chart. Consider the data and answer the accompanying questions.

Number of Miles Driven Chart

Time (in minutes)	Miles Driven
0	0
3	1.5
6	3
9	4.5
12	6

1. How far does Lindsay live from the store?

2. Without graphing, examine the data in the chart, and make a hypothesis about what a scatterplot of this data would look like. Explain your reasoning.

3. Create a scatterplot of the data in the Number of Miles Driven Chart, using a ruler to help you.

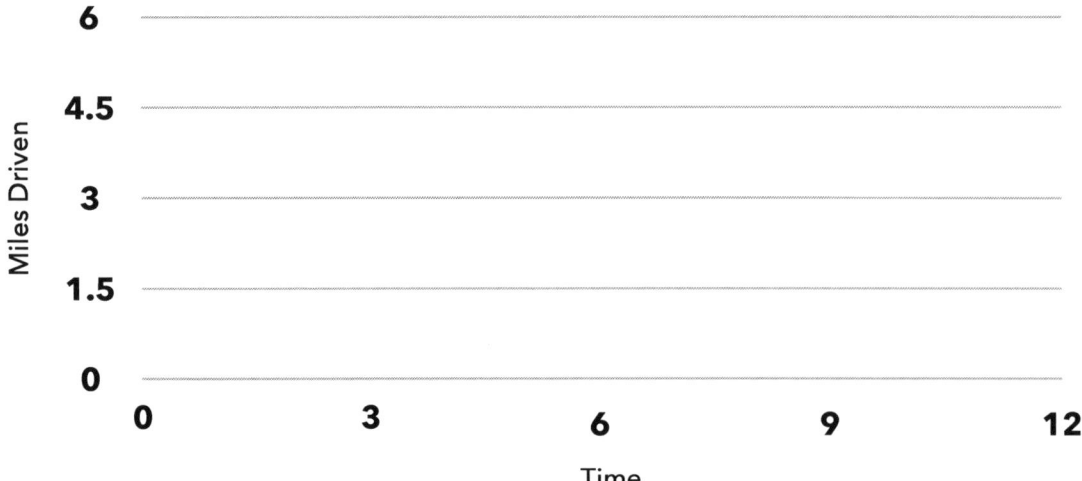

Number of Miles Driven Graph

NAME: _____ DATE: _____

Graphing Time and Distance, continued

4. What is the shape of the graph?

5. Was your hypothesis correct?

6. This graph is linear. Explain a situation in which the graph of time versus miles driven might not be linear.

7. If Lindsay kept driving at the same rate of speed, how many miles would she have traveled after 15 minutes? How far would she have traveled after one hour?

8. If Lindsay kept driving at the same rate of speed, how long would it take her to travel 15 miles?

LESSON 5.2
APPLY NONLINEAR DATA, CHARTS, AND GRAPHS

RESOURCES AND MATERIALS

- Lesson 5.2 Number Lines and Patterns
- Lesson 5.2 Graphing Temperature Data
- Student math notebooks
- Two glasses or cups (per group of four); one with warm but not boiling water, the other with cool water, chilled with ice
- A thermometer (per group of two)
- Ruler and/or measuring tape (per group of two)
- Stopwatch
- Graphing calculators and/or computers with Microsoft Excel

ESTIMATED TIME

60 minutes

OBJECTIVES

In this lesson, students will:
- analyze real-life, nonlinear data,
- create a chart and graph for the data collected, and
- interpret the meaning of the data collected.

PRIOR KNOWLEDGE

Students should know how to measure temperature and how to chart and graph data.

Unit 5: Graphing, Data, and Charts in Algebra

INSTRUCTIONAL SEQUENCE

Anticipatory Set (10 minutes): Have students complete Lesson 5.2 Number Lines and Patterns. Students may work in groups or alone. Lead a discussion about the questions.

Activity (45 minutes):
1. Explain to students the difference between a linear graph and a nonlinear graph. A linear graph will have y values increase or decrease at equal rates for equal-spaced values of x.
2. Divide students into groups of four to begin. Within each group of four, divide the students into two groups of two. One group of two will have a glass of cool water, and the other group of two will have a glass of warm water.
3. Instruct students to place a thermometer in each glass of water. Set a timer for 2 minutes. Once the thermometers have been in the water for 2 minutes, say "measure." The students should record the temperature in the appropriate chart on Lesson 5.2 Graphing Temperature Data.
4. Continue to call out "measure" every 2 minutes over a period of 10 minutes as students record the data in the chart provided.
5. After 10 minutes, instruct the groups of two to get back together with their groups of four and share the information they recorded.
6. Students should discuss any patterns they see and then answer the questions on Lesson 5.2 Graphing Temperature Data.
7. While the students wait for each 2-minute time interval to be called out in order to measure the water temperature, they should conduct another set of measurements: Using a ruler, have students measure each other's height, the length of their hands from wrist to middle fingertip, and the length of their feet. Students should record the data collected in their math notebooks.
8. Create a class chart, using data from all of the students.
9. Using the data from the chart, students should graph height versus hand length on one graph (this may be created on a graphing calculator using the scatterplot graph, or in an Excel spreadsheet). Students will then sketch the graph in their math notebooks. Students should graph height versus foot length on a second graph, and then sketch this graph in their notebooks.
10. Discuss the following questions as a class:
    - Do the points on the graph look like they would form a straight line when connected?
    - Is there a point that looks like it may not fit (an outlier) on either of the graphs? What happens if that point is discarded?
    - Does the graph look like a linear graph? If the graph is not linear, what shape does it resemble?
    - Explain any patterns you see in the graph.

THINKING LIKE A MATHEMATICIAN

Exit Ticket (5 minutes): Have students answer the following in their math notebooks: *Did the warm water and cool water change temperature at the same rate, or did one change temperature faster than the other? Explain how you know this.*

EXTENSION ACTIVITY

Have students respond to the following in their math notebooks:
- Explain a real-life scenario in which a graph would be linear. How do you know that?
- Explain a real-life scenario in which a graph would be nonlinear. How do you know that?

ASSESSMENT OBSERVATIONS

- Students will work together to determine that the data are not changing in a linear fashion. (The data points and graph suggest a nonlinear change for both the cool and warm water.)
- Students will compare and contrast the growth rate or decline rate for the water temperature with the other group members. They should include information about the graph being nonlinear.

NAME: _____ DATE: _____

LESSON 5.2
Number Lines and Patterns

Directions: Complete the following questions about number lines and patterns.

1. Put a point on the following number line for each number given: 0, 3, 6, 9

2. Explain the pattern you find in the numbers.

3. How are the numbers spaced on the number line?

4. Put a point on the following number line for each number given: 0, 1, 4, 9

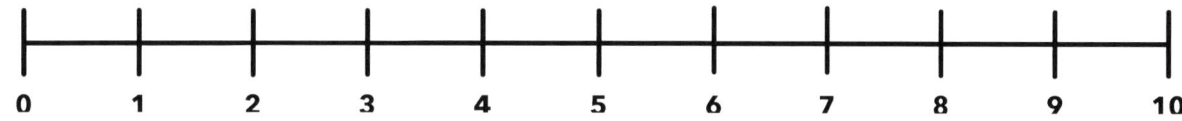

5. Explain the pattern you find in the numbers.

6. How are the numbers spaced on the number line?

Thinking Like a Mathematician © Prufrock Press Inc.

NAME: _____ DATE: _____

LESSON 5.2
Graphing Temperature Data

Directions: Fill in the chart with the time and temperature for the cool or warm water glass. Then answer the questions that follow.

Water Chart

Time (minutes)	Temperature (degrees)
0 minutes	
2 minutes	
4 minutes	
6 minutes	
8 minutes	
10 minutes	

1. Will the data in the chart result in a graph of a linear or nonlinear shape?

2. How do you know that?

3. Graph the points on a graphing calculator. Explain the shape of the graph.

LESSON 5.3

PROJECT: CREATE A BUSINESS PLAN

RESOURCES AND MATERIALS

- Lesson 1.3 Reflection
- Lesson 5.3 Business Plan Project
- Lesson 5.3 Rubric
- (Optional) Student computers with graphing software, such as Microsoft Excel

ESTIMATED TIME

120 minutes

OBJECTIVES

In this lesson, students will:
- design a small business,
- develop a business plan, and
- analyze cost versus income.

PRIOR KNOWLEDGE

Students should know how to read and analyze graphs, compare values, and explain basic ideas of economics (e.g., cost and income).

INSTRUCTIONAL SEQUENCE

1. Model how to design a small business, using Lesson 5.3 Business Plan Project as a guide. Model for students how to determine costs, set a product price, and calculate the profit margin.

THINKING LIKE A MATHEMATICIAN

2. Help students decide what good or service to market for their own small business. They may work in small groups or individually. They will need to develop a business plan, including the cost and estimated income for each good or service they provide, using Lesson 5.3 Business Plan Project.
3. Have students develop a presentation to share their business. See Lesson 5.3 Rubric for presentation criteria.

TEACHER NOTES

In this project, students will design a small business. They could consider operating a lawn-mowing business or offering goods like jewelry. This crosscurricular, open-ended project gives you flexibility to extend students' thinking in ways that are meaningful in your classroom.

Each student's graph (see Lesson 5.3 Business Plan Project) will look a little different, depending on the type of business each student chooses to develop. A sample completed graph could look like Figure 4.

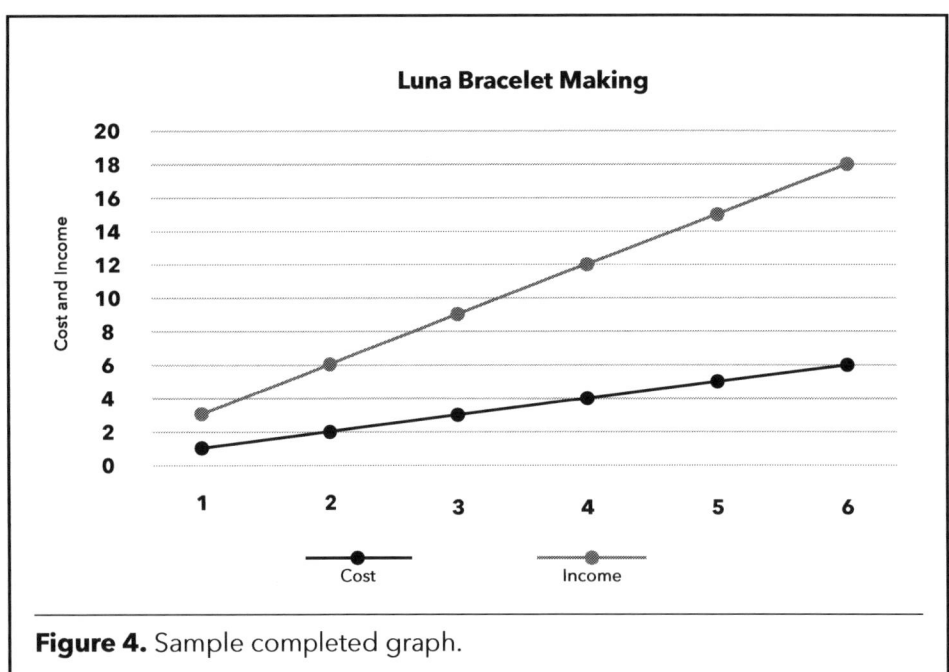

Figure 4. Sample completed graph.

Notice that both the costs and income are on the same graph. A successful business will have the income line greater than the cost, indicating a profit. Graphing both the cost and the income on one graph will help students observe both elements together, and your discussion can help them understand why small businesses need to consider both elements carefully.

Graphing software like Microsoft Excel can help students tailor their individual graphs into a professional-looking design. However, students can also tailor the blank template included in this project.

Unit 5: Graphing, Data, and Charts in Algebra

EXTENSION ACTIVITIES

- Connect students with a local newspaper to discuss marketing opportunities for their small businesses.
- Invite a small business owner to speak to the class about budgeting. Invite a city manager to discuss challenges and opportunities for business in your area.
- Have students read about other people who have started similar businesses. Your school librarian can help develop a list of fiction and nonfiction books on economics.

NAME: _____ DATE: _____

LESSON 5.3
Business Plan Project

Directions: Design a business plan. Consider the cost of operation as well as the income to determine your profit.

1. Name of your business: _____

2. What good or service does your business provide?

3. What costs will you need to pay? Include costs like:
   - equipment/materials,
   - marketing, and
   - salary for employees.

Item	Cost
Total Operating Cost	

NAME: _____ DATE: _____

Business Plan Project, continued

4. Determine the total cost of each item you sell. Show your equation below. Then use a calculator to solve it.
    - If you are creating a product, calculate it like this: Cost = total expenses ÷ number of products.
    - If you are providing a service, calculate it like this: Cost = total expenses ÷ number of hours of work.

    Equation: _____

    Solution: Each item costs _____ to create.

5. Fill in Data Table #1 below. Your cost should be a pattern.

    Data Table #1

Number of Items	Cost to Create

6. Now think about your income. Consider how much it costs to create each item, as well as how much your customers are willing to pay. How much will you charge per item? _____

NAME: _____ DATE: _____

Business Plan Project, continued

7. Fill in Data Table #2 about your income.

   Data Table #2

Number of Items	Cost to Create

8. Use the information from your data tables to build a graph:
   - Label the horizontal axis with the number of products.
   - Label the vertical axis with the dollar values.
   - You will draw two lines on the graph: one showing the cost and the other showing your income.

   Use information from Data Table #1 for the cost. Use your pencil to chart this information first. When you're sure about your information, use green to draw the line.

   Use information from Data Table #2 for the income. Use your pencil to chart this information first. When you're sure about your information, use blue to draw the line.

Business Name:

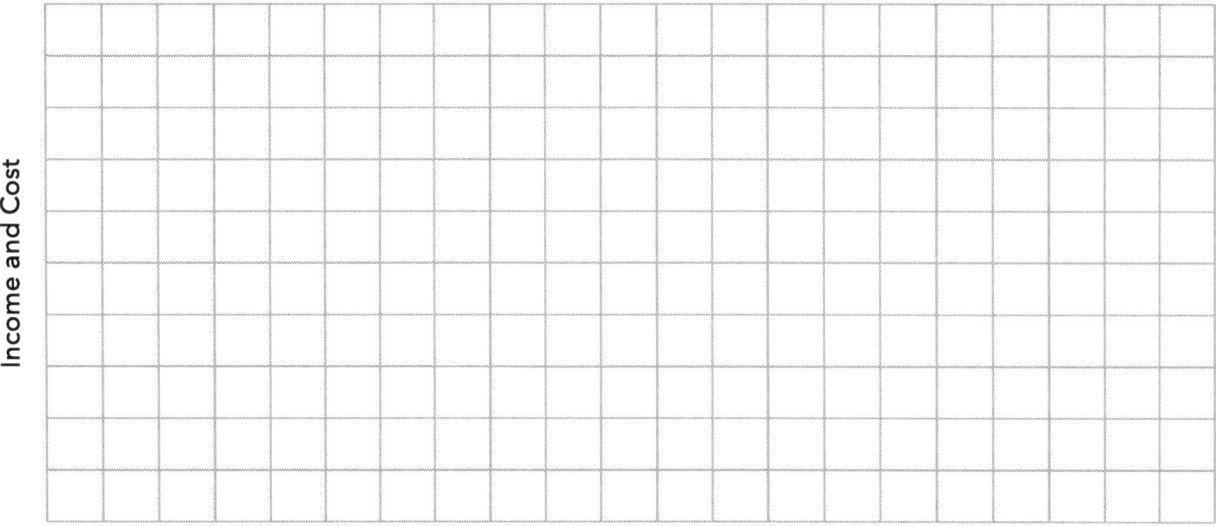

Number of Items Sold

NAME: _____ DATE: _____

LESSON 5.3 RUBRIC
Business Plan Project

Criteria	Points Possible	Points Earned
Business Plan Development		
Student's written explanation is neat, and letters are properly formed.	10	
Student's writing includes proper grammar and spelling.	20	
Student used proper capitalization and punctuation.	20	
Student included math vocabulary.	20	
Student included details about the product, cost, income, and profit.	30	
Total:	100	
Graph		
Title, x-axis, and y-axis are properly labeled.	20	
Student's graph accurately reflects costs of the small business, with at least five data points.	20	
Student's graph accurately reflects income of the small business, with at least five data points.	20	
Student's graph is neat and easy to read.	20	
Student can explain the data on the graph.	20	
Total:	100	
Presentation Grade		
Student presented project professionally, standing with tall posture.	20	
Student spoke loudly and clearly.	20	
Student explained essential aspects of the small business with detail: • Business plan. • Costs of business. • Income from business.	30	
Student displayed the graph in a way that the audience could view.	10	
Student responded to audience questions with authority.	10	
Student was a thoughtful audience member.	10	
Total:	100	
Work Ethic and Reflection		
Student's work was completed and submitted on time.	25	
Student's work was neat, tidy, and presentable.	25	
Student thoughtfully reflected on work throughout the process.	25	
Student incorporated feedback in meaningful ways.	25	
Total:	100	

UNIT 6
GEOMETRY

RATIONALE

Geometry is one of the five NCTM content standards. Students should have a basic understanding of quadrilaterals in geometry, including certain properties for specific types of quadrilaterals. This unit allows students to discover some of those properties and make mathematical arguments about geometric properties.

PLAN

In Lesson 6.1, students will explore geometric concepts of perimeter and area by constructing and flying a kite. In Lesson 6.2, students will complete an application activity to discover properties of special quadrilaterals. In Lesson 6.3, students will explore famous architectural creations around the world, analyze them, and then share their discoveries by creating a visual presentation.

LESSON 6.1

EXPLORE PERIMETER AND AREA

RESOURCES AND MATERIALS

- Lesson 6.1 Measuring Your Kite
- Lesson 6.1 Exit Ticket
- Two 48-inch dowels (per group)
  - One dowel marked at 24 inches
  - One dowel marked at 16 inches

- String (large spool for each group)
- Newspaper or large roll of paper
- Scissors (per group)
- Tape (per group)

ESTIMATED TIME

75–85 minutes

OBJECTIVES

In this lesson, students will:
- create a kite using dowels, string, paper, and tape;
- predict, test, and analyze the flight of each kite; and
- discover how the perimeter and area may affect the flight of the kite.

PRIOR KNOWLEDGE

Students should have basic knowledge of triangles, perimeter, and area. They should understand perpendicular lines and right angles.

Unit 6: Geometry

INSTRUCTIONAL SEQUENCE

Anticipatory Set (5 minutes): Have students consider the following in their math notebooks: *What shapes are used to create kites? What shapes have you seen in kites? What makes a kite fly?* Share answers in a whole-class discussion.

Activity (60–70 minutes):
1. Place students in small groups of 3–4. Each group should receive two marked dowels (see Resources and Materials), string, tape, scissors, and paper.
2. Instruct students to attach the dowels at the markings at a 90-degree angle using string and tape. Students should tie the string to a dowel first, and then wrap the string around the point of intersection, making an "x" (see Figure 5, Step 1). They should then tape the string in place.

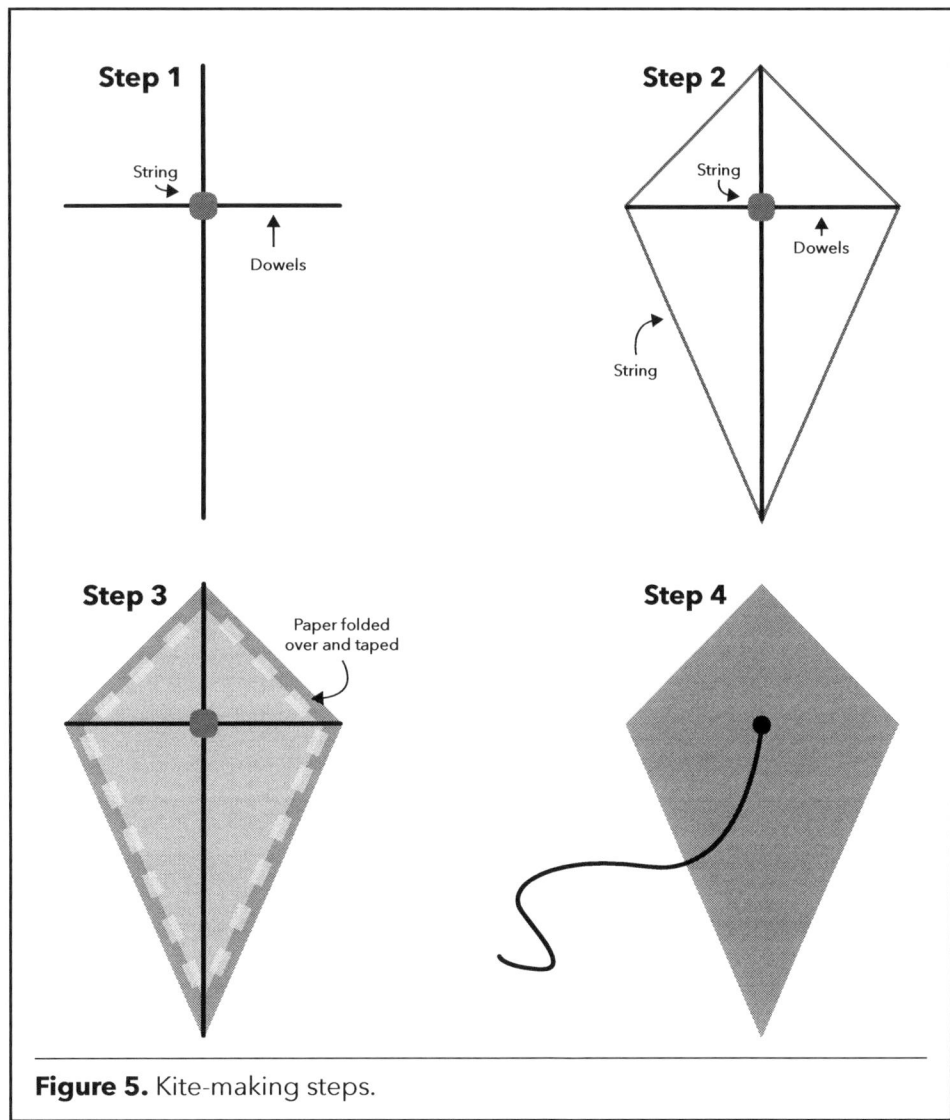

Figure 5. Kite-making steps.

THINKING LIKE A MATHEMATICIAN

3. Students should use another string to outline the shape of a kite, wrapping and taping the string at the end of each dowel (see Figure 5, Step 2).
4. Once students have finished the outline of their kites, lead a discussion about the shapes formed in the kites (e.g., the kite is a quadrilateral and is created with triangles; the dowels separate the kite into four right triangles; the triangles are congruent; the kite has symmetry).
5. Instruct students to add the paper. They should fold the edges of the paper about 2 inches over the string and tape them to the underside of the kite (see Figure 5, Step 3).
6. Students should then poke a small hole in the paper where the dowels intersect and are tied together. Instruct them to thread a new string (about 5–6 feet) through the hole (from the solid paper side of the kite) and tie the string to the dowels. They should make sure the string is tied tightly. The bulk of the string should come out on the side that does not show the dowels (see Figure 5, Step 4).
7. Distribute Lesson 6.1 Measuring Your Kite and have students answer the first two questions about their kites.
8. Then, lead a class discussion about the perimeter, area, and the shape of the kites. How might those components affect the flight of the kites? Discuss how to calculate perimeter and area.
9. Have students calculate the perimeter and area of their kite to complete the rest of the handout.
10. Once students have finished, you may take the class outside to fly the kites.

Exit Ticket (10 minutes): Distribute copies of Lesson 6.1 Exit Ticket and have students record their impressions.

EXTENSION ACTIVITY

Have students complete the following questions about kites.
- Do you think a change in the perimeter of a kite will change the way the kite flies? Why or why not?
- Do you think a change in the area of a kite will change the way the kite flies? Why or why not?
- Do you think a change in the shape of a kite will change the way the kite flies? Why or why not?

ASSESSMENT OBSERVATION

The students should be able to find the perimeter and area of the kite they designed.

NAME: _____ DATE: _____

LESSON 6.1
Measuring Your Kite

Directions: Use a ruler to measure the dimensions of your kite and determine the perimeter and area.

1. Measure the length of each side of your kite. Label the measurements on the picture of the kite below.

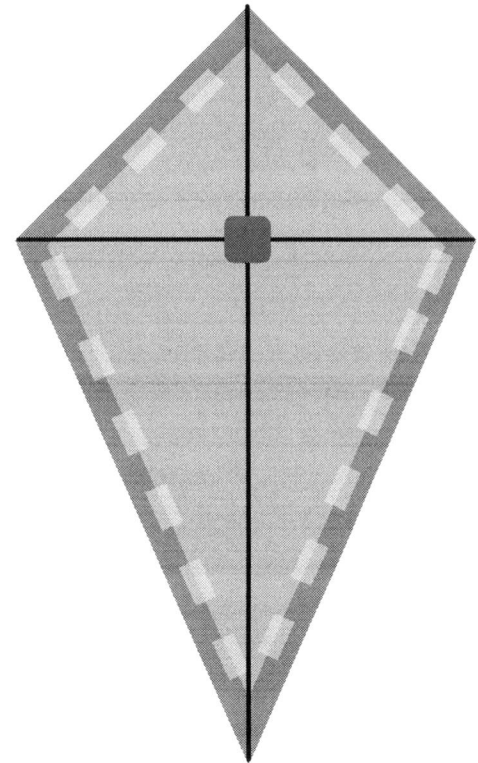

2. Find the perimeter of your kite.

3. Find the area of your kite in square inches. Remember, your kite is made up of two triangles. Show your work below.

Thinking Like a Mathematician © Prufrock Press Inc.

101

NAME: _____ DATE: _____

LESSON 6.1
Exit Ticket

Directions: Consider your calculations and your kite's flight as you answer the following questions.

1. Explain how you calculated the perimeter of your kite. Which material used in building the kite represented the perimeter of the kite?

2. Explain how you calculated the area of your kite. Which material used in building the kite represented the area of the kite?

3. What changes would you make the next time you build a kite to improve the flight?

LESSON 6.2
APPLY PROPERTIES OF QUADRILATERALS

RESOURCES AND MATERIALS

- Lesson 6.2 Measuring Special Quadrilaterals
- Student math notebooks
- Graph paper
- Ruler (per student)
- Protractor (per student)

ESTIMATED TIME

60–70 minutes

OBJECTIVES

In this lesson, students will:
- construct a rectangle, draw diagonals, and discuss any patterns they notice;
- measures the diagonals of special quadrilaterals; and
- predict and discover the properties of special quadrilaterals.

PRIOR KNOWLEDGE

Students should have knowledge of:
- special quadrilaterals (e.g., parallelogram, rectangle, rhombus, square, and trapezoid),
- how to measure in centimeters, and
- how to use a protractor to measure angles.

THINKING LIKE A MATHEMATICIAN

INSTRUCTIONAL SEQUENCE

Anticipatory Set (5–10 minutes): Have students use graph paper to graph a rectangle with a base of 3 units and a height of 5 units. They should then find the perimeter and area of the rectangle. Say: *Connect the vertices (corners) of the rectangle. What do you notice?*

Activity (45–50 minutes):
1. Distribute Lesson 6.2 Measuring Special Quadrilaterals. Students should measure the lengths of each part of the diagonals of the parallelograms and fill in the parallelogram chart.
2. Ask students for their measurements and fill in a class chart.
3. Lead a class discussion about the patterns the students found in the parallelograms.
4. Repeat steps 1–3 for the rectangles and trapezoids on the handout.
5. Discuss with the class the properties of parallelograms, rectangles, and trapezoids.
6. Ask students to create their own quadrilateral design on graph paper.
7. Have students reflect on any properties for their own design. Ask: *What properties will always be true?*

Exit Ticket (10 minutes): Have students answer the following questions in their math notebooks:
1. What do you know about the diagonals of parallelograms?
2. Explain what is true for the diagonals of a rectangle that is not true for the diagonals of a trapezoid. Do you think this will always be true?

EXTENSION ACTIVITIES

- Have students use a protractor to measure the angles formed in the parallelograms, rectangles, and trapezoids given in Lesson 6.2 Measuring Special Quadrilaterals.
- Ask students to create a rule that is true for every parallelogram, every rectangle, and every trapezoid, respectively.

ASSESSMENT OBSERVATIONS

Students should begin to see patterns within the special quadrilaterals:
- In parallelograms, opposite sides are congruent (equal measure), and the diagonals bisect (cut into two pieces of equal measure) each other.
- In rectangles, the opposite sides are congruent, the diagonals bisect, and the diagonals are congruent.
- Not all trapezoids have opposite sides congruent, and the diagonals do not bisect each other; only in an isosceles trapezoid will the diagonals be congruent.

NAME: _____ DATE: _____

LESSON 6.2
Measuring Special Quadrilaterals

Directions: Using a ruler, measure the sides and diagonals of the following quadrilaterals in centimeters. Look for patterns as you measure.

Parallelograms

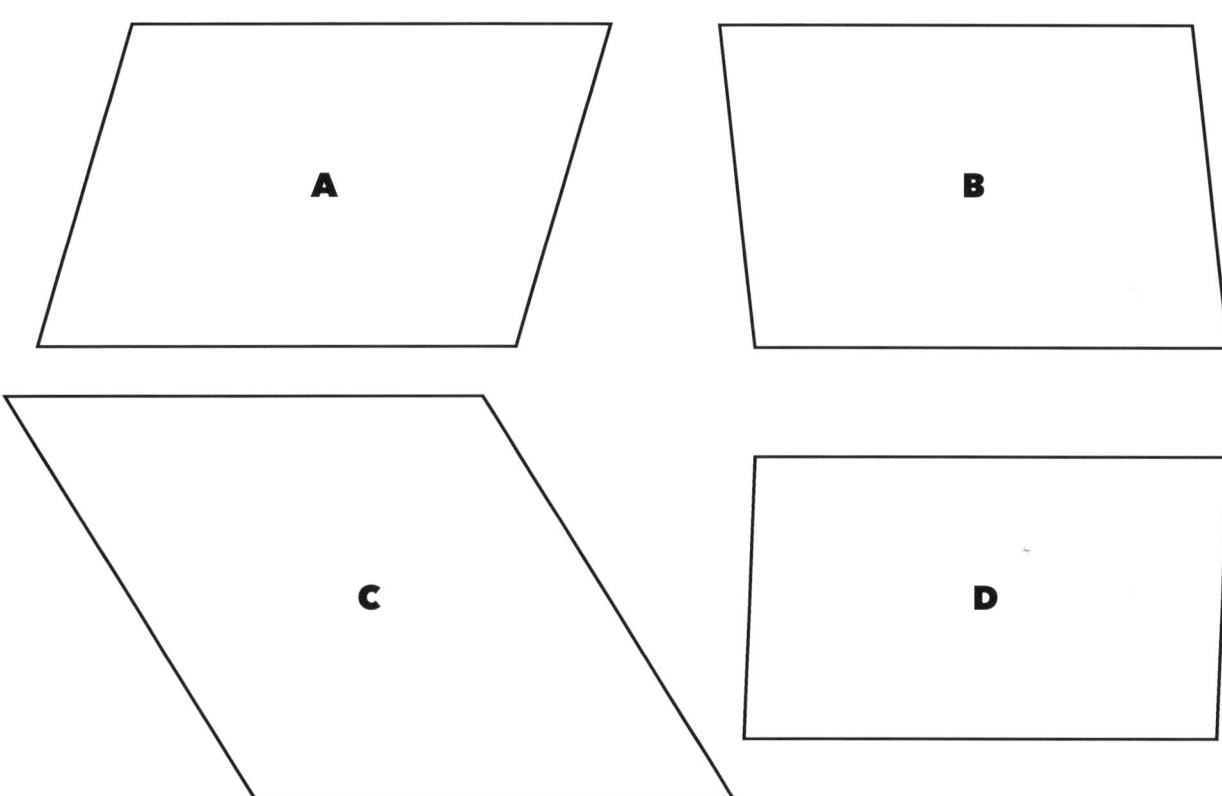

Measure the lengths of the parts of the parallelograms, and fill in the chart.

Parallelogram	Left Side	Right Side	Top	Bottom	Top Left Part of the Diagonal	Bottom Right Part of the Diagonal	Top Right Part of the Diagonal	Bottom Left Part of the Diagonal
A								
B								
C								
D								

What patterns do you see?

Thinking Like a Mathematician © Prufrock Press Inc.

105

NAME: _____ DATE: _____

Measuring Special Quadrilaterals, continued

Rectangles

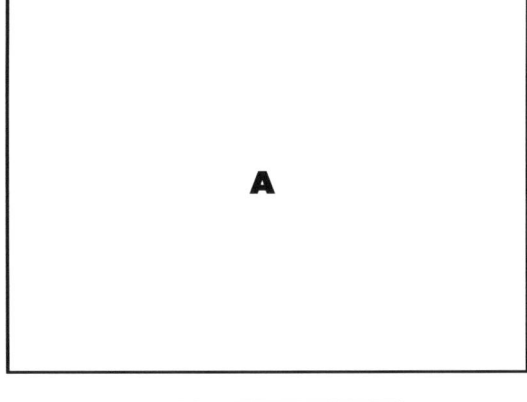
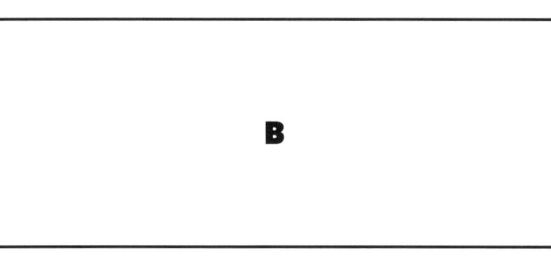
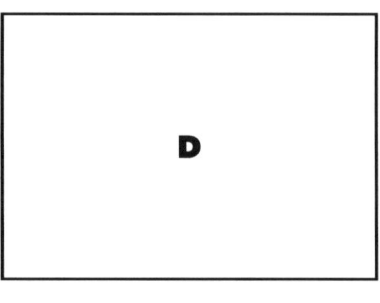

Measure the lengths of the parts of the rectangles, and fill in the chart.

Rectangle	Left Side	Right Side	Top	Bottom	Top Left Part of the Diagonal	Bottom Right Part of the Diagonal	Top Right Part of the Diagonal	Bottom Left Part of the Diagonal
A								
B								
C								
D								

What patterns do you see?

NAME: _____   DATE: _____

Measuring Special Quadrilaterals, continued

Trapezoids

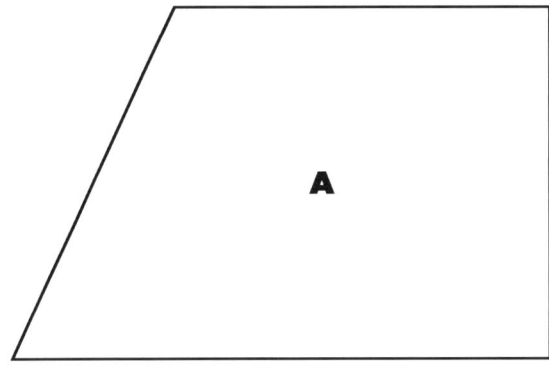
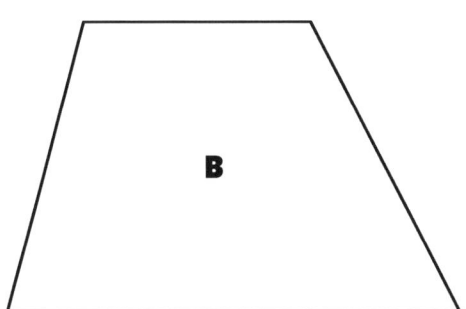
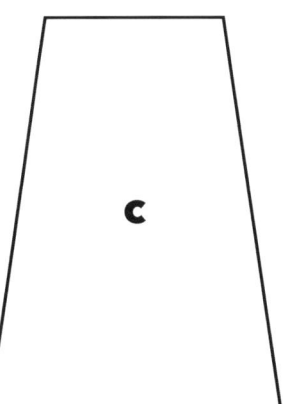
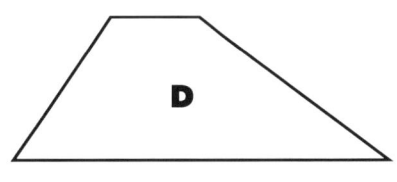

Measure the lengths of the parts of the trapezoids, and fill in the chart.

Trapezoid	Left Side	Right Side	Top	Bottom	Top Left Part of the Diagonal	Bottom Right Part of the Diagonal	Top Right Part of the Diagonal	Bottom Left Part of the Diagonal
A								
B								
C								
D								

What patterns do you see?

LESSON 6.3

PROJECT: ANALYZE FAMOUS ARCHITECTURE

RESOURCES AND MATERIALS

- Lesson 1.3 Reflection
- Lesson 6.3 Architecture Project
- Lesson 6.3 Rubric
- Student computers or tablets with Internet access

ESTIMATED TIME

Will vary based on student engagement

OBJECTIVES

In this lesson, students will:
- research famous architectural structures around the world,
- analyze the geometry of the architecture, and
- create a visual presentation to share their discoveries.

PRIOR KNOWLEDGE

Students should have a working knowledge of two- and three-dimensional shapes, as well as lines.

INSTRUCTIONAL SEQUENCE

1. Show students pictures of your favorite piece of architecture. Describe what makes it so special and the intricate designs that make it visually appealing.

Unit 6: Geometry

2. Introduce the project. Tell students that they will choose a famous architectural structure (see Teacher Notes for ideas) and then analyze the architecture, seeking examples of two- and three-dimensional shapes in addition to lines.
3. Using Lesson 6.3 Architecture Project, students will research and analyze their structures, learning about the location, surrounding area, and geometry of their constructions.
4. Then, have students create a written product to demonstrate their research and knowledge. Guide them to make a meaningful product. They could choose to make a slide show, write an essay, create a diagram with written details, or create another agreeable artifact.
5. Have students present their products to the class.

TEACHER NOTES

Architects employ beautiful shapes and lines to create buildings, bridges, gardens, roadways, and statues. The creations range from deceptively simple to meticulously complex. Your students will benefit the most if they have access to digital pictures of their architecture as well as paper copies where they can write or draw as they analyze geometrical shapes.

The following are possible architectural masterpieces for students to study:
- Burj Khalifa—Dubai, United Arab Emirates
- Sydney Opera House—Sydney, Australia
- Great Wall of China
- Great Pyramid—Giza, Egypt
- Itsukushima Shrine—Miyajima, Japan
- Empire State Building—New York, NY, United States
- Roman Colosseum—Rome, Italy
- Sultan Ahmed Mosque—Istanbul, Turkey
- St. Mark's Cathedral—Venice, Italy
- Taj Mahal—Agra, India
- Saint Basil's Cathedral—Moscow, Russia
- Haeinsa Buddhist Temple—South Gyeongsang Province, South Korea
- Pha That Luang—Vientiane, Laos
- Djenné Mosque—Djenné, Mali
- Guggenheim Museum—New York, NY, United States
- Parthenon—Athens, Greece
- Angkor Wat—Angkor Thom, Cambodia
- Borobudur Temple—Central Java, Indonesia
- The Alhambra—Granada, Spain
- Machu Picchu—Andes Mountains, Peru
- Eiffel Tower—Paris, France
- Big Ben at the Houses of Parliament—London, England
- Golden Gate Bridge—San Francisco, California
- Millau Viaduct—Millau, France
- Rialto Bridge—Venice, Italy

THINKING LIKE A MATHEMATICIAN

- Tower Bridge—London, England
- Chengyang Bridge—Dong Minority Region, China
- Alcántara Bridge—Alcántara, Spain
- Si-o-se Pol (Bridge of 33 Arches)—Isfhan, Iran
- Ponte Vecchio—Florence, Italy
- Pearl Bridge—Kobe-Naruto, Japan
- Pont du Gard Aqueduct—Gard, France
- Seri Wawasan Bridge—Putrajaya, Malaysia

There are several meaningful ways for students to present their research, especially if you choose to use technology. Choose a presentation format that meets the needs of your classroom, considering the amount of time available and the students' opportunities for growth. Their presentations should include an opportunity to demonstrate the geometry in the structures as well as a written component. Whether they write a report on their full research or simply describe it, students should be well prepared with talking points on the essential information.

EXTENSION ACTIVITY

Consider collaborating with experts in your area. A local architect could speak with the students about deliberations that architects need to make, or a city planner could explain how new projects need to fit with preexisting construction. These experts can guide the students to consider practical uses for buildings and bridges, in addition to how the structures make statements.

NAME: _____ DATE: _____

LESSON 6.3
Architecture Project

Directions: Research and analyze your architecture and become an expert! Explore several different resources to help you understand your construction. Remember, research is like a treasure hunt. You may find information in different places and in a different order than is presented in this research guide.

Your famous architecture: _____

Basic Information

1. Who designed your structure? _____

2. When was it built? _____

3. What is the purpose of the structure?

4. How tall is it? _____ meters = _____ feet

5. What materials were used? (Circle all that apply.)

concrete	stone	steel
brick	clay	Other: _____
glass	stained glass	Other: _____

Location

1. On which continent is your structure located? _____

2. In which country is it located? _____

3. In which city or region is it located? _____

NAME: _____ DATE: _____

Architecture Project, continued

4. Indicate the location of your structure on the world map.

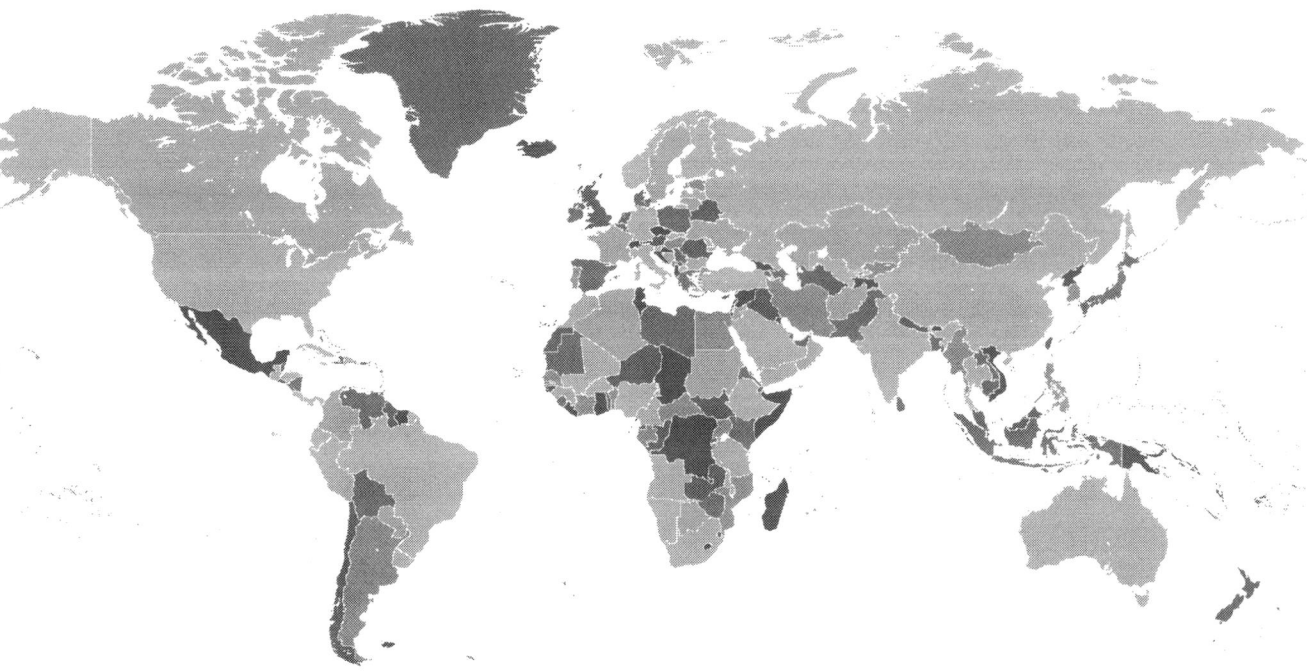

Surrounding Area

1. What is around your structure? (Circle all that apply.)

city	flowers	nature
sculptures	bushes	forest
sidewalks	fountain	river
roads	trees	

2. Other observations about the setting:

3. Look at other structures in the same area to answer the following.

   • Is your structure a similar height as the others?   YES   NO

   • Is it made of similar materials as the others?   YES   NO

NAME: _____ DATE: _____

Architecture Project, continued

- Does it look like it fits in? YES NO
- Does it respect the natural environment? YES NO
- Add details to support your opinions.

Shapes and Lines

1. Observe the two-dimensional shapes on your structure. Take detailed notes so that you can explain your ideas later.

Shape	Where do you observe it?
circle	
triangle	
square	
other quadrilaterals	
pentagon	
hexagon	
decagon	

NAME: _____ DATE: _____

Architecture Project, continued

2. Observe the three-dimensional shapes on your structure. Take detailed notes so that you can explain your ideas later.

Shape	Where do you observe it?
cube	
triangular prism	
rectangular prism	
cone	
pyramid	
cylinder	
sphere	

3. Observe the lines used on your structure. Take detailed notes so that you can explain your ideas later.

Types of Lines	Where do you observe them?
intersecting lines	
perpendicular lines	
parallel lines	
line of symmetry	

UNIT 7
DATA ANALYSIS AND STATISTICS

RATIONALE

Statistics is one of the NCTM content standards. Students will analyze data and apply statistical measures of the center of data. Students will evaluate statistics and use the data to make predictions.

PLAN

In Lesson 7.1, students will use data analysis in this unit to compare different measures of the center of data. In Lesson 7.2, students will analyze tide charts and consider how the phases of the moon affect tide data. In Lesson 7.3, students will explore shipwrecks around the world and learn how the tides impact their visibility.

LESSON 7.1

EXPLORE TIDAL DATA AND STATISTICS

RESOURCES AND MATERIALS

- Lesson 7.1 Tide Measurements
- Student math notebooks
- Student computers or tablets with Internet access and a spreadsheet program
- Graphic calculators (per group)

ESTIMATED TIME

65–80 minutes

OBJECTIVES

In this lesson, students will:
- research, interpret, and graph tide data;
- compare and contrast summary statistics; and
- determine which statistic best represents the tide data.

PRIOR KNOWLEDGE

Students should have experience researching on the Internet. They should also know how to add and divide multidigit numbers.

Unit 7: Data Analysis and Statistics

INSTRUCTIONAL SEQUENCE

Anticipatory Set (10 minutes): Lead a discussion with the class: *Explain what you know about tides. Do tide levels stay the same, or do they change? What causes tides to change?*

Activity (45–60 minutes, depending on research time):
1. Divide students into groups of 2–4 and distribute Lesson 7.1 Tide Measurements.
2. Have groups choose a body of water that has tides. Students will then gather online data of a 30-day period of both high tide levels and low tide levels for their body of water.
3. Instruct students to compile the data on a spreadsheet, computer program, or graphing calculator. They should use the data from the high tide table to create a graph of the day of the month versus the height of the high tide. Then, they can use the data from the low tide table to create a graph of the day of the month versus the height of the low tide.
4. Students should choose which graphical display is best for the data (they may use a scatterplot, a bar graph, or another representation). However, they should use the same type of graph for both high and low tide graphs.
5. When the students have finished their graphs, lead a discussion about finding mean, median, and mode.
6. Have students find the mean, median, and mode of their high and low tide heights, and then record their results on Lesson 7.1 Tide Measurements. Students should decide which statistic best describes their set of data.

Exit Ticket (10 minutes): Ask students to respond to the following in their math notebooks: *Write a paragraph explaining the statistics you chose to represent the high and low tide data (mean, median, or mode). In the paragraph, explain why you chose those statistics as the best representations of the data.*

EXTENSION ACTIVITY

Ask students to compare and contrast the mean, median, and mode of their tide data. They should write a paragraph explaining the similarities and differences in the mean, median, and mode.

ASSESSMENT OBSERVATION

Check for understanding by evaluating exit tickets. Students may have different answers for the statistic they chose to use, but the explanation should contain an understanding of mean, median, or mode, as well as a justification for their choice.

NAME: _____ DATE: _____

Lesson 7.1
Tide Measurements

Directions: You will research to find high tide and low tide measurements from a 30-day period. Record the data in the following tables.

High Tide Measurements

Day	Tide Height	Day	Tide Height	Day	Tide Height
1		11		21	
2		12		22	
3		13		23	
4		14		24	
5		15		25	
6		16		26	
7		17		27	
8		18		28	
9		19		29	
10		20		30	

Mean =

Median =

Mode =

Circle the statistic (mean, median, or mode) that you feel best represents the tide data. Explain why you chose this statistic.

NAME: _____ DATE: _____

Tide Measurements, continued

Low Tide Measurements

Day	Tide Height	Day	Tide Height	Day	Tide Height
1		11		21	
2		12		22	
3		13		23	
4		14		24	
5		15		25	
6		16		26	
7		17		27	
8		18		28	
9		19		29	
10		20		30	

Mean =

Median =

Mode =

Circle the statistic (mean, median, or mode) that you feel best represents the tide data. Explain why you chose this statistic.

LESSON 7.2
APPLY DATA ANALYSIS

RESOURCES AND MATERIALS

- Lesson 7.2 Tide and Moon Phase Data
- Student computers or tablets with Internet access and a spreadsheet application

ESTIMATED TIME

140–145 minutes; this activity may be broken up over several days

OBJECTIVES

In this lesson, students will:
- evaluate and interpret tide and moon phase data,
- predict how the moon phase affects the tide, and
- compare and contrast hypotheses created by student groups.

PRIOR KNOWLEDGE

Students should have an understanding of the phases of the moon, as well as high tide versus low tide (see Lesson 7.1).

INSTRUCTIONAL SEQUENCE

Anticipatory Set (10–15 minutes): Explain the phases of the moon to students or have students research on their own. Students could record explanations and pictures of the moon's phases in their math notebooks. Ask students to recall what they learned in Lesson 7.1: *Explain the difference between high tide and low tide.*

Activity (120 minutes, time will vary based on research time):
1. Divide students into groups of two and distribute Lesson 7.2 Tide and Moon Data. They will use the data they compiled in Lesson 7.1 for the high tides, and then research the phases of the moon for that same 30-day period.
2. Have students compare the moon data with the tide data.
3. Have students give the phases of the moon numeric values: new moon = 1, waxing crescent = 2, first quarter = 3, waxing gibbous = 4, full moon = 5, waning gibbous = 6, last quarter = 7, waning crescent = 8.
4. Have students use a spreadsheet, graphing calculator, or computer program to create a graph of the tide data. The graph should include moon phases as a part of the x-axis, and tidal measurement as part of the y-axis.
5. Ask students to make a hypothesis about the relationship between the phases of the moon and the tides.
6. Students should compare their hypothesis with another group's hypothesis.
7. Lead a class discussion about the hypotheses the students made. What are the similarities? What are the differences?

Exit Ticket (10 minutes): Have students research when the full moon and the new moon will appear next. Students should use that data to make a prediction about the high and low tides on the day of the full moon and the day of the new moon.

EXTENSION ACTIVITY

Have students research an area that experiences high and low tides. They should then create a table that with the best and worst times to scuba dive in that area. Tide and moon data should be used to justify the table created. Students can present the table with justification as a multimedia presentation.

ASSESSMENT OBSERVATIONS

- Students should be able to make predictions based on the information gathered in Lesson 7.2 Tide and Moon Phase Data.
- Students should evaluate patterns in the tide and moon phase data.

NAME: _____ DATE: _____

LESSON 7.2
Tide and Moon Phase Data

Directions: Research tide data and the phases of the moon for a 30-day period. Then fill out the data tables that follow.

High/Low Tide Data

Day	High Tide	Low Tide	Day	High Tide	Low Tide	Day	High Tide	Low Tide
1			11			21		
2			12			22		
3			13			23		
4			14			24		
5			15			25		
6			16			26		
7			17			27		
8			18			28		
9			19			29		
10			20			30		

Thinking Like a Mathematician © Prufrock Press Inc.

NAME: _____ DATE: _____

Tide and Moon Phase Data, continued

Phases of the Moon

Day	Moon Phase	Day	Moon Phase	Day	Moon Phase
1		11		21	
2		12		22	
3		13		23	
4		14		24	
5		15		25	
6		16		26	
7		17		27	
8		18		28	
9		19		29	
10		20		30	

1. Compare your tide data with the moon phase data for the same days. What patterns do you notice?

2. Make a prediction based on the tide and moon data.

Thinking Like a Mathematician © Prufrock Press Inc.

LESSON 7.3

PROJECT: SHIPWRECKED!

RESOURCES AND MATERIALS

- Lesson 1.3 Reflection
- Lesson 7.3 Shipwreck Project
- Lesson 7.3 Rubric

ESTIMATED TIME

Five class sessions of 45 minutes or more, depending on student interest

OBJECTIVES

In this lesson, students will:
- research a famous shipwreck, and
- use local data to investigate how the tide impacts the ship's visibility.

PRIOR KNOWLEDGE

Students should understand that water levels change based on the phase of the moon and the rotation of the Earth (see Lesson 7.2).

INSTRUCTIONAL SEQUENCE

1. Introduce famous shipwrecks by showing pictures of each. History.com (https://www.history.com) and National Geographic (https://www.nationalgeographic.com) are good places to start.
2. Ask the class to brainstorm the many ways that ships may be damaged and sink.

Unit 7: Data Analysis and Statistics

3. Model how to research a shipwreck using Lesson 7.3 Shipwreck Project. Tell students: *Ships rely heavily on data analysis to travel safely around the world. In this project you will investigate a shipwreck along a coast and use local data to understand how much of the wreck is visible during high and low tides.*
4. Guide students to do their own research on their chosen shipwreck, using Lesson 7.3 Shipwreck Project to guide them.
5. Model how to build an infographic (see Teacher Notes). Students should choose about seven or more ideas from their research to include, and they should organize the information in eye-catching ways.
6. Guide students to plan and execute their presentations (see Lesson 7.3 Rubric for presentation criteria).

TEACHER NOTES

Shipwrecks plague each continent. Your students could investigate a wreck in your region or research a wreck from another part of the world. Help them find a wreck with enough public information that students can actually complete the project; many shipwrecks may not have enough information to fully answer students' questions. For example students would find plenty of research about Costa Concordia, a highly publicized wreck off the coast of Italy, but might struggle finding the exact locations, sizes, and other information about lesser-known vessels. Help students recognize that they must look at multiple resources while researching and may still not find everything they need.

Here are some famous shipwrecks:
- Costa Concordia—Italy
- Crested Eagle —Beaches of Dunkirk, France
- Peter Iredale—Fort Stevens State Park, OR
- SS Ayrfield—Homebush Bay, Australia
- Helvetia—Rhossili Beach, Wales
- Queen Anne's Revenge—Atlantic Beach, NC
- USS Arizona—Pearl Harbor, HI
- Astron—Punta Cana, Dominican Republic
- SS Pointe Reyes—Inverness, CA

Students can share their research in a variety of formats, but try introducing them to the idea of an infographic. In this genre, students represent their research in a digital, poster-style format. There are numerous templates available online, or you can ask them to draw an old-fashioned poster. Try websites such as Piktochart (https://piktochart.com), Canva (https://www.canva.com), or Venngage (https://venngage.com), or collaborate with your school's technology or art teachers.

NAME: _____ DATE: _____

LESSON 7.3
Shipwreck Project

Directions: Research a famous shipwreck. Find as much data about the wreck as possible. (*Note.* You may not find all of the information in one place, or the information may not be fully available.) Then, based on your data, make recommendations to the captain about how he or she could have kept the ship safely at sea.

1. Name of your shipwreck: _____

2. What was the purpose of the ship?

3. How long is your vessel?

4. How tall is it?

5. How much did it weigh?

6. Where did it depart from?

7. Where was it going?

8. On what date did it wreck?

NAME: _____ DATE: _____

Shipwreck Project, continued

9. What happened to cause the wreck?

10. Where is it located now?

11. What is the longitude?

12. What is the latitude?

13. Is it still in the water?   YES   NO

14. How deep is it in the water?

15. Is there anything valuable on the ship?   YES   NO

16. What is the estimated value?

17. How many passengers were on board?

18. How many crew members were on board?

19. What other interesting facts did you discover?

NAME: _____ DATE: _____

LESSON 7.3
Rubric

Criteria	Points Possible	Points Earned
Research		
Student's written research is neat, and letters are properly formed.	20	
Student used time wisely.	20	
Student found most answers to research questions.	20	
Student referenced a variety of resources for research.	20	
Student included an appropriate level of detail.	20	
Total:	100	
Infographic		
Student included seven or more pieces of data.	20	
Information reflects thorough, accurate research.	20	
Student's explanation of mathematical concepts is clear and accurate.	20	
Student's work is neat and easy to read.	20	
Student can explain the infographic.	20	
Total:	100	
Presentation Grade		
Student presented project professionally, standing with tall posture.	20	
Student spoke loudly and clearly.	20	
Student explained essential aspects of shipwreck with detail, especially how mathematical concepts apply.	30	
Student displayed the infographic in a way that the audience could view.	10	
Student responded to audience questions with authority.	10	
Student was a thoughtful audience member.	10	
Total:	100	
Work Ethic and Reflection		
Student's work was completed and submitted on time.	25	
Student's work was neat, tidy, and presentable.	25	
Student thoughtfully reflected on work throughout the process.	25	
Student incorporated feedback in meaningful ways.	25	
Total:	100	

FINAL PROJECT
DESIGNING A GARDEN

RATIONALE

Gardens provide rich opportunities to apply many different mathematical concepts into a real-world challenge. Encourage students to consider the topics they explored in the preceding units, such as geometry, number sense, measurement, and patterns, as they design a garden.

PLAN

Students will explore famous and local gardens through pictures or by visiting those places. They will research and design a new mathematically based garden and share their garden designs.

LESSON: FINAL PROJECT
GARDEN MATH

RESOURCES AND MATERIALS

- Lesson 1.3 Reflection
- Final Project Garden Plan
- Final Project Garden Design
- Final Project Rubric
- Gardening books, seed catalogues, and/or books about famous gardens
- Paper and colored pencils/markers

ESTIMATED TIME

Five class sessions of 45 minutes each; may vary depending on student interest

OBJECTIVES

In this lesson, students will:
- research weather patterns and local flora, and
- use mathematical knowledge to design and present a plan for a garden.

PRIOR KNOWLEDGE

Students should understand that different plants have different requirements that must be met in order to help them thrive. Students should also have had some exposure to gardening in different contexts, like rooftop gardens or rural farming.

INSTRUCTIONAL SEQUENCE

1. Explain to students that there are a variety of gardens with different purposes, ranging from decorative gardens to productive food gardens. Explore pictures of each type of garden.

Final Project: Designing a Garden

2. Model the project, selecting your own type of garden and doing research on the variety of plants that are options for the garden. Modeling the process of looking through books and thinking aloud in front of students allows them to see how to navigate obstacles.
3. Give students an opportunity to explore a variety of gardens, either through online opportunities or field trips.
4. Have students research plants and weather patterns for their gardens, using Final Project Garden Plan to guide their research. They may choose to work alone or in groups.
5. Give students time to design their selected type of garden, carefully guiding the research on appropriate types of plants and landscaping decisions, like rock placement. Ask them to consider mathematical concepts, such as patterns, height, or the amount of sunshine required, to help make the most purposeful design possible. Have students complete Final Project Garden Design to aid them in their designs.
6. Have students present their gardens.

TEACHER NOTES

As a culminating project, students will design and present a plan for a garden based on your local geography. This project is highly flexible, designed to fit in any region where you live. For example, rural gardeners may have large acreage available, gardens in the suburbs may fit neatly in a corner of your school campus, or urban growers may design rooftop or container gardens. This can be a prime opportunity for your class to visit local gardens, or you may invite specialists to speak at the school. The experts can share their own designs, offer advice or feedback, or be guest judges for a culminating class-wide competition.

With an open-ended project like this, be sure to set a few constraints for the sake of time. Students could spend weeks developing every aspect of their plans. You should consider how much time you want to dedicate to the project. Then, consider how much you want to narrow the students' focus.

Students' work should focus on mathematical concepts that are observable in their garden design, whether that involves geometry, measurement, time, or other concepts that they discover.. You may need to model your own design or brainstorm a class list of concepts that apply to horticulture. For example:

- Measure your garden in terms of length, width, area, perimeter, cubic feet, and water requirements.
- What shape(s) is the garden bed? What angles and geometry can you identify?
- Design a blueprint of the plants, and consider the timing of when to plant the seeds. Identify the length of time they take to grow and the harvest time involved.
- Manage a budget for the garden. How much do the startup supplies cost? What opportunities for profit are involved?
- How is counting involved? Can you use your garden as an opportunity to teach younger children to count?

NAME: _____ DATE: _____

FINAL PROJECT
Garden Plan

Directions: Sketch a design of your garden. You will use this plan to create a final draft. Your plan should include:

1. A title
2. A map key, showing types of plants, water resources, buildings, fences, and important features of your garden
3. Map scale
4. Compass rose

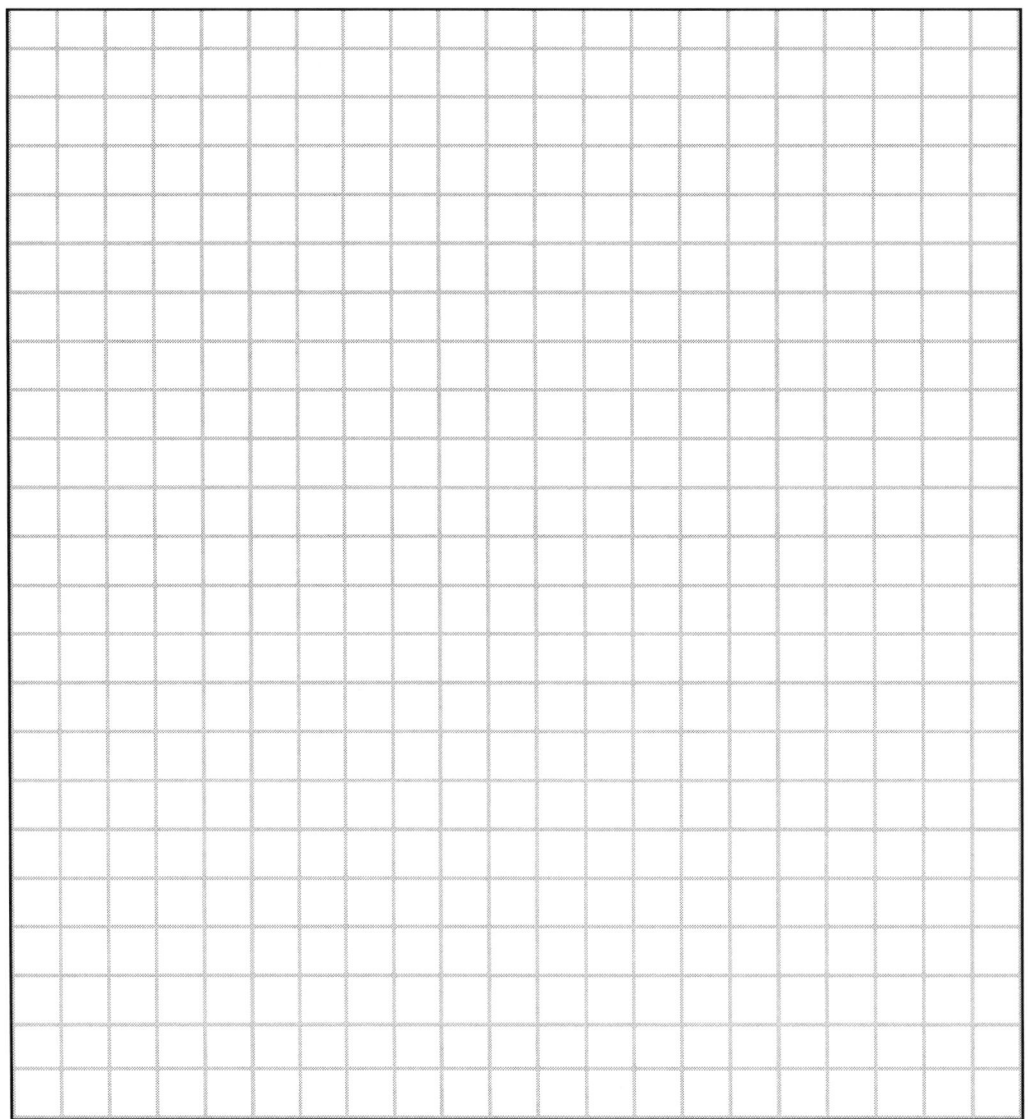

NAME: _____ DATE: _____

Garden Plan, continued

Explain your garden from three math perspectives.

Math concept #1: _____

How does this math apply to your garden?

Math concept #2: _____

How does this math apply to your garden?

Math concept #3: _____

How does this math apply to your garden?

NAME: _____ DATE: _____

FINAL PROJECT

Rubric

Criteria	Points Possible	Points Earned
Garden Plan Development		
Student's written explanation is neat, and letters are properly formed.	10	
Student's writing includes proper grammar and spelling.	20	
Student used proper capitalization and punctuation.	20	
Student documented the resources used for research.	10	
Student included math vocabulary.	10	
Student explained the math concepts in detail, including a thorough description and appropriate vocabulary.	30	
Total:	100	
Formal Map and Math Concepts		
Title and compass rose are properly labeled.	10	
Student's plan accurately reflects garden measurements to scale.	10	
Map key is accurate.	10	
Student's explanation of three mathematical concepts is clear and accurate.	30	
Student's work is neat and easy to read.	20	
Student can explain the plan.	20	
Total:	100	
Presentation Grade		
Student presented project professionally, standing with tall posture.	20	
Student spoke loudly and clearly.	20	
Student explained essential aspects of garden with detail, especially how three mathematical concepts apply.	30	
Student displayed the map in a way that the audience could view.	10	
Student responded to audience questions with authority.	10	
Student was a thoughtful audience member.	10	
Total:		
Work Ethic and Reflection		
Student's work was completed and submitted on time.	25	
Student's work was neat, tidy, and presentable.	25	
Student thoughtfully reflected on work throughout the process.	25	
Student incorporated feedback in meaningful ways.	25	
Total:	100	

ANSWER KEY

UNIT 1

LESSON 1.1 BASE-4 COUNTING

000	001	002	003
010	011	012	013
020	021	022	023
030	031	032	033
100	101	102	103
110	111	112	113
120	121	122	123
130	131	132	133
200	201	202	203
210	211	212	213
220	221	222	223
230	231	232	233
300	301	302	303
310	311	312	313
320	321	322	323
330	331	332	333

THINKING LIKE A MATHEMATICIAN

1. Answers will vary.
2. Answers will vary.
3. Answers will vary.
4. You add a number to the tens digit.
5. You add a number to the hundreds digit.

LESSON 1.1 EXTENSION ACTIVITY

110 + 011	223 + 121
121	1010

323 − 122	301 − 223
201	12

LESSON 1.2 ROBOT MAZE

Program code:
1. Turn counterclockwise 90 degrees.
2. Move forward one block.
3. Turn clockwise 90 degrees.
4. Move forward one block.
5. Turn counterclockwise 90 degrees.
6. Move forward one block.
7. Turn counterclockwise 90 degrees.
8. Move forward one block.
9. Turn clockwise 90 degrees.
10. Move forward one block.
11. Turn clockwise 90 degrees.
12. Move forward two blocks.

LESSON 1.2 WRITING IN BINARY CODE

1. 0000, 0001, 0010, 0011, 0100, 0101, 0110, 0111, 1000, 1001, 1010
2. Answers will vary.
3. Answers will vary.
4. Answers will vary.

LESSON 1.2 EXTENSION ACTIVITY

- 011, 1010, 1100
- 010, 001, 011, 101

Answer Key

UNIT 2

LESSON 2.1 CALCULATING ORDER OF OPERATIONS

1. Answers will vary.
2. 11.5 (four-function calculator) or 21 (graphing/scientific calculator)
3. Answers will vary.
4. Answers will vary. Sample answer: The first calculator worked all of the operations from left to right. The second calculator used the order of operations.
5. Answers will vary.
6. 21
7. The graphing/scientific calculator.
8. Answers will vary. Sample answer: The graphing calculator used the order of operations.

LESSON 2.1 EXIT TICKET

1. $3 + 2 \times 5 - 10 \div 2 = 8$
2. $12 \div 4 - 6 \times 1 + 10 - 3 = 4$
3. $24 - 6 \div 2 \times 3 - 10 + 5 \div 5 = 6$
4. $5 - 5 + 2 \div 1 \times 3 = 6$

LESSON 2.2 PARENTHESES IN ORDER OF OPERATIONS

1. 31
2. What is inside parentheses must occur first.
3. Answers will vary
4. a. 3; b. 26
5. Answers will vary

LESSON 2.2 EXPONENTS IN ORDER OF OPERATIONS

1. 20
2. Answers will vary. Sample answer: Exponents will happen before multiplication and division.
3. Answers will vary.
4. a. 2; b. 69
5. Answers will vary.

THINKING LIKE A MATHEMATICIAN

LESSON 2.2 EXIT TICKET

1. 18
2. 27
3. 9
4. Answers will vary. Sample answer: Order of operations may be used to solve multistep problems, and if order of operations is not used correctly, the problem is incorrect.

LESSON 2.2 EXTENSION ACTIVITIES

1. 15
2. 37
3. 42

LESSON 2.3 EQUATION CARDS

5	15
13	20
12	100
29	3
28	-4

UNIT 3

LESSON 3.1 ANTICIPATORY SET

1. 10, 12
2.

LESSON 3.1 SEQUENCES

1. Answers will vary.
2. Answers will vary. Sample answer: Adding three to the term before.
3. 13, 16
4. a. Answers will vary; b. Answers will vary
5. Answers will vary. Sample answer: The terms are getting smaller. The terms are all even.

Answer Key

6. Answers will vary. Sample answer: Dividing the previous term by 2.
7. 4, 2
8. Answers will vary. Sample answer: There are two ones. The numbers increase.
9. 21, 34, 55, 89, 144
10. Answers will vary. Sample answer: The two numbers before are added to get the next number.
11. Answers will vary.

LESSON 3.1 EXTENSION ACTIVITY

- Answers will vary. Sample answer: 3n – 2
- Answers will vary.

LESSON 3.2 ANTICIPATORY SET

1, 1, 2, 3, 5, 8, 13, 21, 24, 55

LESSON 3.2 THE GOLDEN SPIRAL

1. The sides of the squares are the lengths in the Fibonacci sequence.
2. Answers will vary. Sample answer: By connecting the vertices of each square.
3. Answers will vary. Sample answers: A seashell, our galaxy, a pinecone.

LESSON 3.2 EXTENSION ACTIVITIES

The ten quotients are: 1, 2, 1.5, $1.\overline{6}$, 1.6, 1.625, 1.61538462, 1.61904762, 1.61764706, $1.6\overline{18}$

UNIT 4

LESSON 4.1 ANTICIPATORY SET

1. 24 hours
2. 60 minutes
3. 1,440 minutes
4. The time the Earth takes to rotate once on its axis.
5. The time the Earth takes to revolve once around the sun.
6. No. Sample answer: Planets rotate on their axes in different amounts of time. Planets revolve around the sun at different rates. Planets have different lengths of orbits around the sun. It takes some planets longer to orbit the sun than others.

THINKING LIKE A MATHEMATICIAN

UNIT 5

LESSON 5.1 ANTICIPATORY SET

1. We measure the speed of a car in miles per hour.
2. We may use inches, centimeters, feet, miles, kilometers, etc.
3. We may use seconds, minutes, hours, days, years, decades, etc.

LESSON 5.1 GRAPHING TIME AND DISTANCE

1. 6 miles
2. Answers will vary. Sample answers: The time and distance are both increasing. Lindsay drives half the miles as the number of minutes.
3. Graph the data from the Number of Miles Driven Chart. Students should plot the points first, and then connect them when they discover that the graph is linear.

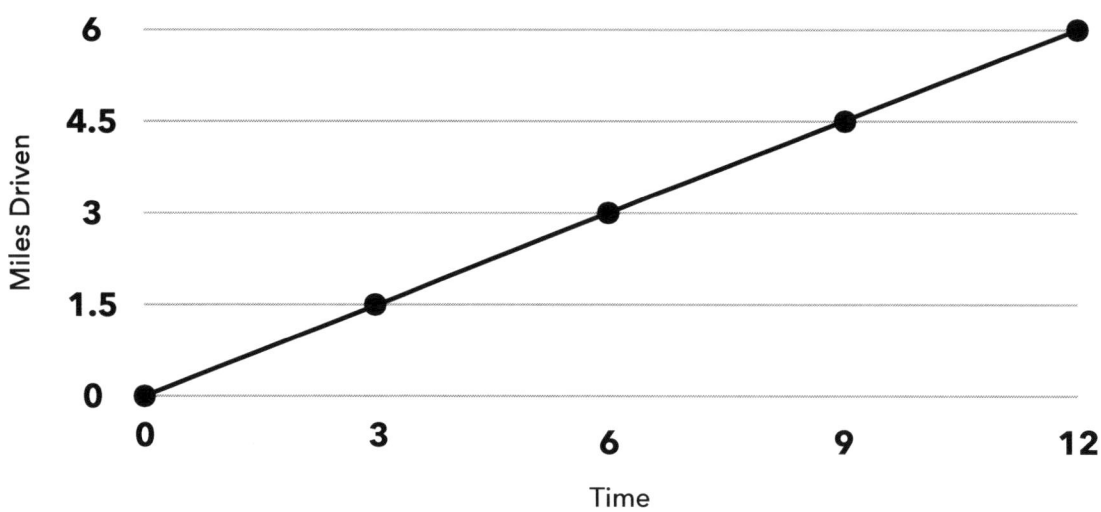

Number of Miles Driven Graph

4. If the points are connected, the graph is a straight line.
5. Answers will vary.
6. Answers will vary. Sample answer: If the speed limit changes, you might not have a line.
7. 7.5 miles; 30 miles
8. It would take her 30 minutes.

Answer Key

LESSON 5.1 EXTENSION ACTIVITY

1. 30 miles per hour
2. 3 minutes
3. Yes, Lindsay is speeding on her way home. If she drives double the speed to the store, she is driving home at 60 miles per hour.

LESSON 5.2 NUMBER LINES AND PATTERNS

1.

2. Each number increases by three. The points are equally spaced.
3. Answers will vary. Sample answer: There are points every three units.
4.

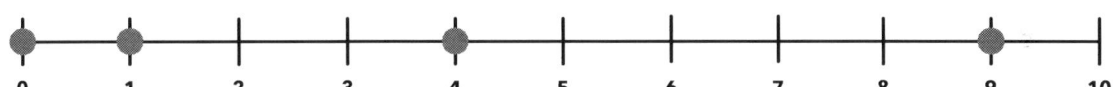

5. Answers will vary. Sample answer: The distance between points increases by 1, then 3, then 5. The numbers are perfect squares.
6. Answers will vary. Sample answer: They are spaced farther apart as you go along the axis.

LESSON 5.2 GRAPHING TEMPERATURE DATA

1. Nonlinear
2. Answers will vary. Sample answer: The temperature differences are not the same at each time measurement.
3. Answers will vary.

UNIT 6

LESSON 6.1 MEASURING YOUR KITE

1. The top two sides measure about 28.9 inches each, and the bottom two sides measure 40 inches each. The picture of the kite should be labeled with the correct measurements. The measurements will depend on the placement of the string and may vary by several inches.
2. 137.7 inches
3. 1152 square inches

LESSON 6.1 EXIT TICKET

1. Add the sides together to get the perimeter. The string outside represents the perimeter.
2. The kite is made up of two (or four) triangles. We calculate area using the triangle formula for area (A = 1/2 × base × height). The paper represents the area of the kite.
3. Answers will vary. Possible answers include: Change the dimensions, use different paper, use different string, vary the materials.

LESSON 6.2 MEASURING SPECIAL QUADRILATERALS

Students should draw the diagonals for each quadrilateral using a ruler, and then measure all of the diagonals and sides. Exact measurements will vary depending on the scale and printer used to reproduce the handout. You may instruct students to measure to the nearest centimeter if needed.

Parallelograms

Answers will vary. Sample answers: The opposite sides are the same. The diagonals cut each other in half. One diagonal is longer than the other. The top left and bottom right diagonals are the same measure, and the top right and bottom left diagonals are the same measure.

Rectangles

Answers will vary. Sample answers: The opposite sides are the same measure. The parts of the diagonals are all the same measure.

Trapezoids

Answers will vary. Sample answers: The sides aren't always the same. The diagonal parts are not always the same measure.

LESSON 6.2 EXIT TICKET

1. Answers will vary. Sample answer: The parts of the diagonals are the same length.
2. Answers will vary. Sample answer: Parts of the diagonals of the parallelograms are the same length, but that is not true of the diagonals of a trapezoid.

ABOUT THE AUTHORS

Mary-Lyons Walk Hanks holds a BS in mathematics from Virginia Tech and an MS in secondary education from Old Dominion University. She is a math teacher at Lafayette High School in Williamsburg, VA. She also serves as a clinical faculty member for the William & Mary Teacher Preparation Program. She lives in Williamsburg, VA, with her husband and daughter.

Jennifer K. Lampert, a graduate of Roanoke College, teaches third grade at The Matthew Whaley School in Williamsburg, VA. She also serves as a Clinical Faculty member for the William & Mary Teacher Preparation Program. She lives in Williamsburg with her husband, two sons, and basset hound.

Katherine Plum holds an M.A.Ed. from William & Mary and is a National Board Certified Teacher. She teaches third grade at Matthew Whaley School in Williamsburg, VA. She lives with her husband, daughter, twin sons, and two rescue dogs.